Valuing Oil Spill Prevention

THE ECONOMICS OF NON-MARKET GOODS AND RESOURCES

VOLUME 5

Series Editor: Dr. Ian J. Bateman

Dr. Ian J. Bateman is Professor of Environmental Economics at the School of Environmental Sciences, University of East Anglia (UEA) and directs the research theme Innovation in Decision Support (Tools and Methods) within the Programme on Environmental Decision Making (PEDM) at the Centre for Social and Economic Research on the Global Environment (CSERGE), UEA. The PEDM is funded by the UK Economic and Social Research Council. Professor Bateman is also a member of the Centre for the Economic and Behavioural Analysis of Risk and Decision (CEBARD) at UEA and Executive Editor of Environmental and Resource Economics, an international journal published in cooperation with the European Association of Environmental and Resource Economists (EAERE).

Aims and Scope

The volumes which comprise *The Economics of Non-Market Goods and Resources* series have been specially commissioned to bring a new perspective to the greatest economic challenge facing society in the 21st Century; the successful incorporation of non-market goods within economic decision making. Only by addressing the complexity of the underlying issues raised by such a task can society hope to redirect global economies onto paths of sustainable development. To this end the series combines and contrasts perspectives from environmental, ecological and resource economics and contains a variety of volumes which will appeal to students, researchers, and decision makers at a range of expertise levels. The series will initially address two themes, the first examining the ways in which economists assess the value of nonmarket goods, the second looking at approaches to the sustainable use and management of such goods. These will be supplemented with further texts examining the fundamental theoretical and applied problems raised by public good decision making.

For further information about the series and how to order, please visit our Website
http:/www.wkap.nl/series.htm/ENGO

Valuing Oil Spill Prevention:
A Case Study of California's Central Coast

Richard T. Carson
University of California, San Diego

Michael B. Conaway
University of Alabama

W. Michael Hanemann
University of California, Berkeley

Jon A. Krosnick
Ohio State University

Robert C. Mitchell
Clark University

Stanley Presser
University of Maryland

KLUWER ACADEMIC PUBLISHERS
DORDRECHT / BOSTON / NEW YORK

A C.I.P. Catalogue record for this book is available from the Library of Congress.

ISBN 0-7923-6497-X
ISBN 1-4020-2864-4 (eBook)

Additional material to this book can be downloaded from http://extras.springer.com

Published by Kluwer Academic Publishers
PO Box 17, 3300 AA Dordrecht, The Netherlands

Sold and distributed in North, Central and South America
by Kluwer Academic Publishers,
101 Philip Drive, Norwell, MA 02061, U.S.A.

In all other countries, sold and distributed
by Kluwer Academic Publishers,
PO Box 322, 3300 AH Dordrecht, The Netherlands

Printed on acid-free paper

DEDICATION

This monograph is dedicated to Sara Russell
without whose persistence and sorely-tried patience
this study would never have been completed.

Contents

*Some appendices may be found only in the book, only on the CD, or on both. See section 1.5 for a description of the appendices.

Preface

On April 22–23, 1988, approximately 432,000 gallons of San Joaquin Valley crude oil spilled from an aboveground storage tank at a Shell Oil Company refinery into the surrounding environment, including the northern reaches of San Francisco Bay (the Martinez spill). Pursuant to the settlement of the resulting litigation (*United States/California v. Shell Oil Co.*, No. C89-4220, (N.D. Cal 1990)), Shell Oil Company provided funding for, among other things, studies to improve future response strategies to oil spills and ensure better restoration of resources and services affected by such spills. The California Oil Spill Contingent Valuation Study was one of the studies funded by the settlement.

The purpose of the California Oil Spill (COS) Contingent Valuation (CV) Study was "to execute and document a contingent valuation study of natural resource damages associated with offshore, coastal, or estuarine oil spills in California" (State of California, Department of Justice, Contract Number 89-2126). The COS CV study developed an estimate of per household *ex ante* economic value for a program to prevent a specified set of natural resource injuries to those species of birds and intertidal life that are consistently affected by oil spills along California's Central Coast.

The principal investigators on the COS study team were Richard T. Carson of the University of California, San Diego, and W. Michael Hanemann of the University of California, Berkeley. The study's project manager was Kerry M. Martin of Natural Resource Damage Assessment, Inc. Donald R. McCubbin of the University of California, San Diego, provided assistance with various survey design, statistical, and programming issues as did Craig Mohn of the University of California, Berkeley, Nick Flores of the University of Colorado, Boulder (formerly of the University of California, San Diego), and David Chapman of the National Oceanic and Atmospheric Administration (NOAA) (formerly of the University of California, Berkeley). Paul A. Ruud of the University of California, Berkeley and Thomas Wegge of Jones and Stokes Associates also provided support for various aspects of the study.

Westat, Inc. of Rockville, Maryland (Martha Berlin, program manager) administered the pilot surveys and the main survey; and the Survey Research Center, University of Maryland (Johnny Blair, program manager) developed the sample weights. Jones and Stokes Associates and Natural Resource Damage Assessment, Inc. (NRDA) provided administrative and logistical support.

Patricia Wynne designed the saltwater ecosystem graphics. Technical assistance was provided by Larry Espinosa and Pierre duVair of the California Department of Fish and Game (CDFG) and Norman Meade of NOAA. They also served as conduits for review by state and federal scientists and as reviewers in their own right.

The study's internal peer reviewers were Raymond J. Kopp of Resources for the Future and John B. Loomis of Colorado State University, Fort Collins. The study's external peer reviewers were Richard C. Bishop of the University of Wisconsin, Madison, John P. Hoehn of Michigan State University, and Norbert Schwarz of the University of Michigan, Ann Arbor. Bishop and Hoehn are resource economists; Schwarz is a psychologist and survey researcher.

Funding for this project came from several sources. The primary source of funding was the settlement agreement ending litigation over the Martinez spill (Contract between the State of California, Department of Justice, and W. Michael Hanemann, Contract Number 89-2126; Sara Russell, Project Coordinator). Supplemental funding for the COS study was provided by the Office of Oil Spill Prevention and Response (OSPR) of the California Department of Fish and Game (CDFG) (Contract between the State of California, Department of Fish and Game and the University of California, San Diego, Contract Number FG 3499 OS; Pierre duVair, Project Manager) and the Damage Assessment Center of the National Oceanic and Atmospheric Administration (NOAA) (Subcontract between Industrial Economics, Inc. and Westat, Inc., NOAA Contract No. 50-DGNC-1-00007, Task Order No. 56-DGNC-4-50081; Norman Meade, Task Order Manager). Finally, additional support was provided by Natural Resource Damage Assessment, Inc.

Comments or questions may be sent by e-mail to Michael.Conaway@ua.edu; please put "COS-Book" in the subject line.

CHAPTER 1

Introduction and Overview

1.1. Policy Context

Huge oil spills capture the attention of the public and policymakers alike. The recent spill off the coast of Spain involving the tanker *Prestige* engendered active discussion within the European Union about liability rules and the measurement of damages. This same discussion arose in the United States more than ten years earlier when the *Exxon Valdez* tanker spill, the largest tanker spill in U.S. waters, prompted the United States Congress to pass the Oil Pollution Act (OPA) in 1990 to help prevent the pollution of coastal waters and seas by oil.[1] At about the same time, California, the country's largest producer and consumer of petrochemical products, enacted the Lempert-Keene-Seastrand Oil Spill Prevention and Response Act (OSPRA) to help protect the State's 1,000 miles of coastline from oil spills.[2] The potentially devastating environmental and economic impact of oil spills mandates that we discover how much it is worth to prevent oil spills.

California has a long history of oil spills. While the majority of these spills were small, much of the total quantity of oil was released from a few large spills.[3] California's largest oil spill occurred in 1969 when an oil platform blowout in the Santa Barbara Channel spilled two million gallons of crude oil. The much smaller Shell Martinez refinery spill in 1988 resulted in the release of over 430,000 gallons of oil into an area including San Francisco Bay. In 1990, the *American Trader*'s anchor punctured its hull releasing over 390,000 gallons

1 OPA provisions include the establishment of tanker-free zones in environmentally sensitive areas, the use of tug escorts in certain busy tanker lanes, and the requirement that all tankers operating in U.S. waters be equipped with double-hulls by the year 2015. The act also addresses liability issues.

2 OSPRA expanded the authority, responsibility, and duties of the Department of Fish and Game for marine oil spills, emphasizing oil spill prevention as well as contingency planning, enforcement, and response. It also created the Office of Oil Spill Prevention and Response (OSPR) within the California Department of Fish and Game. OSPR works closely with the U.S. Coast Guard, the Department of Commerce's National Oceanic and Atmospheric Administration (NOAA), and the Department of Interior's Fish and Wildlife Service and Minerals Management Service.

3 The U.S. Coast Guard maintains several extensive databases on oil spills occurring in U.S. waters. A 1993 report by Mercer Management Consulting, Inc. for OSPR provides detailed information on California's marine facilities and tank vessel traffic, oil spills, spill clean-up and damage costs, and coastline characteristics.

of oil into the coastal waters off Huntington Beach just south of Los Angeles. In addition to these very large spills, a series of smaller but still significant spills occurred that might have been avoided or the harm from them minimized by taking appropriate action. These spills tend to occur in three areas: the greater Los Angeles area, San Francisco Bay, and California's Central Coast, which lies between Los Angeles and San Francisco. In this study, we focus on moderately large spills along California's Central Coast, a region that includes some of the most scenic and ecologically important areas of the State, including the well-known Big Sur region.

Many policy decisions involving California's Central Coast area depend upon estimates, whether implicit or explicit estimates, of the value to the public of preventing oil spills in particular areas. Such decisions include how far off shore to route oil tanker and barge traffic, the frequency of pipeline inspections, and requirements embodied in oil spill response plans. To determine worth (*i.e.*, monetary value) to the public, economists typically look to information about the public's preferences. The most commonly consulted source of such information is market price. In this instance, however, prevention of the harm caused by oil spills is not a commodity that an individual can readily buy or sell in the marketplace; oil spill prevention is what economists call a non-marketed good. Environmental economists have devoted considerable attention to the issue of how to place a monetary value on non-marketed goods. At present, the most commonly used approach for valuing non-marketed goods is contingent valuation, the approach taken in this study.

1.2. Overview of the Contingent Valuation Method

Contingent valuation (CV) is a sample survey-based, economic methodology that can be used to obtain data from which economic values may be constructed for a wide array of economic goods, including non-marketed goods, such as improved air and water quality.[4] CV has been used for both policy purposes and litigation by numerous state and federal government agencies. For example, CV was used by the U.S. Environmental Protection Agency to value the benefits of the Clean Water Act (Carson and Mitchell, 1993a; U.S. EPA, 1994), and by the State of Alaska in estimating the passive use losses resulting from the *Exxon Valdez* oil spill (Carson *et al.*, 1992; Carson *et al.*, 2003).

S. V. Ciriacy-Wantrup, a professor at the University of California, Berkeley, and one of the first economists to specialize in environmental and natural resource economics, laid out the conceptual foundations of contingent valuation in 1947, although other prominent economists (*e.g.*, Bowen, 1943) had already started to think about the use of surveys to measure the demand for public

[4] *See* Mitchell and Carson (1989) and Bateman *et al.* (2002) for comprehensive reviews of the theoretical and empirical basis of CV. Carson (2000) provides a non-technical overview of CV for policymakers who are considering the use of results from a particular CV study.

goods. The first published academic application waited until 1963 when Robert Davis published a paper in *Natural Resources Journal* that valued outdoor recreation in Maine.[5] Since 1963, the number of published CV studies has grown rapidly, reflecting applications not only valuing environmental goods but also other types of non-marketed goods.[6] A recent bibliography (Carson, forthcoming) lists over 5,000 CV papers and studies from over 100 countries. A large part of the growth in CV can be attributed to its frequent use by U.S. Federal agencies, including the U.S. Environmental Protection Agency, the U.S. Forest Service, the Department of Commerce, and the Department of Interior, and state agencies, as well as by government agencies in a number of other countries and international organizations, such as the World Bank and the Inter-American Development Bank.

The theoretical foundation of CV is the same as that underlying all economic valuation, regardless of whether the valuation is based on market transactions or other non-market techniques (*e.g.*, the travel cost method used to value recreational activities). In all forms of economic valuation, the analyst constructs an economic value from an observed choice and from knowledge of the circumstances of that choice. Unlike other valuation methods, CV gives an analyst control over the choice presented and over the circumstances by which the choice is framed. In contrast, other valuation methods usually rely on recorded past choices, and the analyst must make assumptions about the features of the choice outside his or her knowledge and control. Of the three basic non-market valuation methodologies (Freeman, 1993), hedonic pricing (*e.g.*, property value and wage models), household production function (*e.g.*, travel cost analysis and averting behavior), and CV, CV is the only methodology capable of including passive use value[7] in its estimate of total economic value.

The use of CV has been the subject of an on-going debate in the academic literature and in various policy forums.[8] A recent review of CV by an independent blue ribbon panel convened by NOAA and chaired by Nobel Prize winners Kenneth Arrow and Robert Solow provides recommendations for conducting CV surveys and concludes that "CV studies can produce estimates

[5] The first reported application of this methodology appears to be a 1957 study commissioned by the National Park Service (Audience Research, Inc., 1958).

[6] *See* Hanemann (1992) and Carson (forthcoming) for brief reviews of the history of CV.

[7] The term passive use value was first used in *Ohio v. U.S. Department of the Interior*, 880 F.2d 432 (D.C. Cir. 1989), and is synonymous with or inclusive of a number of other terms that have been used in the economic literature including non-use values, existence values, stewardship values, bequest values, and option values. *See* Carson, Flores, and Mitchell (1999) for a review of the theoretical and empirical issues related to passive use values.

[8] See the series of articles in the American Agricultural Economic Association's journal *Choices* (Carson, Meade, and Smith, 1993; Desvousges *et al.*, 1993; Randall, 1993) and in the American Economic Association's *Journal of Economic Perspectives* (Portney, 1994; Diamond and Hausman, 1994; Hanemann, 1994).

reliable enough to be the starting point for a judicial or administrative determination of natural resource damages – including passive use values" (Arrow *et al.*, 1993).[9]

1.3. Issues Related to Conducting a CV Survey Before a Spill

Contingent valuation surveys conducted for damage assessment purposes have usually been designed and administered after the damage occurred (*e.g.*, Carson *et al.*, 1992; Carson *et al.*, 2003). In contrast, the survey we present here was conducted before any particular oil spill occurred. Administering the CV survey before a spill has several advantages and disadvantages relative to conducting the study after a particular spill. A major advantage is that the government agency that conducts a CV study before the spill can release that study and effectively post its expectation of the magnitude of the monetary damages if a spill occurs. This number can be a valuable input to decision-makers considering public policies and regulatory decisions related to oil spill prevention.

If the cost of the oil spill injuries falls to those best able to prevent oil spills, firms transporting oil, it is important to reduce uncertainty surrounding those costs so that responsible parties will implement the economically appropriate level of prevention. The posting of such numbers can dramatically reduce uncertainty to such firms and their insurance companies over expected damages as measured by a CV survey focusing on a particular set of injuries. This uncertainty has been a major criticism by some critics of the use of the use of CV for damage assessment purposes (*e.g.*, Daum, 1993). NOAA's Blue Ribbon Panel on CV (Arrow *et al.*, 1993) recommended that *reference studies* be undertaken to reduce the uncertainty about the costs of an oil spill. The present study represents the first major effort to develop an approximate damage estimate before any spill injuries takes place.

While it might be argued that all CV studies for damage assessment should be done in advance of an accident, the number of CV studies necessary to cover all feasible accidents would be extremely large and hence prohibitively costly. Indeed, the oil spill valued in any CV study conducted for damage assessment purposes before an actual oil spill will almost surely have a different set of characteristics from the actual spill. Thus, the injuries from a particular

9 The Exxon-sponsored volume edited by Jerry Hausman (1993) serves as a comprehensive source critiquing CV. We address various issues raised by CV critics in a number of recent papers, *e.g.*, Carson (1997, 2000), Carson *et al.* (1997), Carson, Flores, and Hanemann (1998), Carson, Flores *et al.* (1996), Carson, Flores and Meade (2001), Carson, Flores and Mitchell (1999), Carson, Groves and Machina (1999), Carson, Hanemann *et al.* (1998), Carson and Mitchell (1993b; 1995), Flores and Carson (1997), Hanemann (1994; 1995; 1999), Hanemann and Kanninen (1999) and Mitchell and Carson (1995).

In addition, a critique of the study reported on in this volume was funded by an unspecified group of companies and submitted to the government as a regulatory comment (Dumford *et al.*, 1996). Our detailed response to that critique is found in Appendix L.

oil spill for which a damage assessment is desired will likely be larger or smaller than those from the oil spill depicted in the scenario presented in the CV survey along one or more dimensions of the spill, such as the number of birds killed or the number of miles of beach oiled. The objective then should be to choose CV scenarios that are either close to expected accidents or which will help to more tightly bound the expected magnitude of damages from different injury scenarios.

There is an important difficulty though with the valuation of a general *ex ante* spill prevention plan for an area that should be recognized. The CV scenario for preventing another single big spill like the *Exxon Valdez* spill may appear to be much more definitive to some respondents. This is because the actions taken to prevent such a spill are those that directly address the problems that caused the earlier spill and because the injuries from the spill to be prevented can be easily cast as similar to those from the earlier spill. A more generic spill prevention scenario will of necessity depict a set of injuries that are *expected not to occur* by preventing spills in a particular geographic area. From a perspective of making public policy decisions, this more generic nature of spill injuries is probably desirable due to the probabilistic nature of spills and spill prevention efforts. From a litigation perspective, a CV scenario which is close to (but not identical to) the set of injuries that actually occurred may help to contribute to an early settlement by reducing uncertainty over the damage number that would be obtained if a CV study for the specific set of injuries for a spill was undertaken.

The survey instrument for this study incorporates many elements that are likely to be useful features in a wide variety of CV studies that focus on oil spills. The survey instrument was based upon extensive background research and builds upon previous high quality CV survey instruments, particularly that done for the *Exxon Valdez* oil spill (Carson *et al.*, 1992; Carson *et al.*, 2003). Use of this survey as a starting point should reduce survey development time and the risk of presenting a scenario lacking key elements. It should be directly transferable with minor changes for use with other similar coastal oil spill prevention scenarios. Several elements of the survey instrument, *e.g.*, some of the visual aids, should be transferable to an even wider range of oil spill scenarios with little modification.

This study also makes a number of advances with respect to statistical methodology. It further develops the Turnbull estimator,[10] uses a Box-Cox construct validity model, and examines the usefulness of a cluster approach to explain differences in respondent willingness to pay amounts. The CV data

[10] The original proposal for this estimator was by Turnbull (1976) in the biometrics literature. Carson and Steinberg (1990) and Kristrom (1990) first used variants of the estimator in a CV context. Recent contributions include Haab and McConnell (1997; 2002) and Hanemann and Kanninen (1999).

collected is examined using a variety of qualitative techniques as well as quantitative tests that should be useful to other researchers.

Thus this study may serve as a reference not only for bounding the probable damages from an actual spill, but also as a model of best practices. The techniques reported here may serve as a reference for other researchers by providing a model for generating a robust estimate of damages.

Since this study was intended to survey as a *reference* CV study for the government, industry groups financed an extensive critique of the original study (Dunford *et al.*, 1996) and submitted that critique as a comment to several government agencies.[11] In this reply we address a number of key issues related to the general use of contingent valuation for valuing goods, such as oil spills injuries, with extensive passive use considerations. A number of technical issues related to the analysis of contingent valuation data are also addressed. We respond to concerns raised by Dunford *et al.* concerning the design and analysis of this particular study. We believe that this exchange will be useful to those who conduct CV studies and those who are trying to evaluate and potentially use the results of our studies.

Reference CV studies such as this one have a public goods aspect. They are designed to serve a number of distinct purposes without having been done explicitly for any single one. They are not done to provide the damage estimate for a particular spill nor to help answer a specific policy question or to test a distinctly new methodology. As such, we believe that reference CV studies will be underprovided. Regret concerning their relative scarcity will always come after they are needed.

1.4. A Brief Summary of Study

The COS study team designed and implemented a CV survey following best-available practices for survey design and administration. In the survey, respondents were given the opportunity to vote *for* or *against* a government program financed by a one-time income tax surcharge on California households. The program would prevent, over the next decade, natural resource injuries from oil spills that harm wildlife and shoreline along California's Central Coast.

The per household sample estimate of total *ex ante* value obtained from the study is $76.45 (with a standard error of $3.78). The statistical approach used to obtain this estimate is a non-parametric maximum likelihood procedure developed by Turnbull (1976) which yields a lower bound on the sample mean. The estimate includes an adjustment for respondents who did not pay California

[11] Our reply is contained in Appendix L of this book. The reply was written with the hope that it is self-contained in the sense that the Dunford *et al.* (1996) comment is summarized before our response to it is provided. The complete Dunford *et al.* critique is available from Triangle Economic Research (www.ter.com) or from bama.ua.edu/~issr/cosbook.html.

taxes; that adjustment treats the votes of non-taxpaying respondents *for* the program as votes *against* the program.

The CV survey on which this estimate is based is the culmination of an extensive program of instrument development including focus groups, in-depth pretest interviews, and a pilot study with an instrument similar to that used in the final study. For the main survey, Westat, Inc. completed 1,085 in-person interviews with a random sample of English-speaking California households, achieving a response rate of 74.4 percent.

The qualitative and quantitative responses to the main survey were analyzed in order to assess the validity and reliability of the measure of value and, as this measure is constructed from respondents' choices, the meaningfulness of those choices. These analyses support the validity and reliability of the valuation results.

Qualitative survey data provided evidence that respondents paid attention to the survey and took their choice seriously and that their choices reflected their perceptions of and preferences for the program. Furthermore, responses to open-ended questions that asked respondents about their choices suggested a good understanding on the part of the respondent of what the program would accomplish and what the program would cost.

1.5. Organization of Book

The validity and reliability of a survey depends on the quality of its design and administration. The design and administration of the COS CV survey and the analysis of the collected data were guided by many considerations, including those raised in Arrow *et al.*, 1993, those derived from experience with past natural resource damage assessments and past public policy evaluations involving non-marketed goods, and other research conducted by the principal investigators and other members of the study team. Chapter 2 outlines the design and development phases of the survey instrument. Chapter 3 describes the wording, format, and sequence of the final survey instrument. Chapter 4 discusses the administration of the main study survey, including the sample design, interviewer training and supervision, quality control, completion rates, sample weights, and data entry. Chapter 5 evaluates the responses to questions pertaining to respondents' choices and perceptions of the scenario and responses to interviewer-evaluation questions. Chapter 6, the final chapter, presents the statistical framework for the analysis and the quantitative results, including the estimate of total *ex ante* economic value and its sensitivity to alternative ways of treating the data and a construct validity model which relates willingness to pay (WTP) to various respondent characteristics.

A wealth of additional materials is contained in the appendices of the book. Some of these appendices are provided in the book itself; some are provided on the accompanying compact disk (CD); and some in both places.

The heart of any CV study is its survey instrument and accompanying show cards and graphic materials. This study's survey instrument is provided in Appendix A of the book to help aid the interpretation of the results presented. It is also provided on the CD in Adobe PDF format to help facilitate its use in courses. We believe the survey instrument from this study can serve as a valuable starting point for the development of CV survey instruments in other areas.

Appendix B (CD only) provides information on the sample design and execution, including a number of different forms used for this purpose. Appendix C contains a tabulation of the closed-ended questions and is provided on the CD. Appendix D (book only) contains the coding categories for the open-ended questions in the survey instrument. Appendix E (CD only) contains the actual responses to the open-ended questions. Appendix F (book only) contains a mathematically oriented discussion of the main statistical estimator used, the Turnbull lower bound on mean willingness to pay. Appendix G (CD only) contains an extensive set of cross-tabulations suggested by NOAA's Blue Ribbon Panel on CV (Arrow *et al.*, 1993) that examine how the response to the key choice question varied with a long list of variables. Appendix H (book only) provides a set of supplemental analysis tables for Chapter 6. Appendix I (book only) provides a comparative analysis of the results of this study with those from the Carson *et al.* (1992; 2003) *Exxon Valdez* oil spill study. Appendix J (CD only) provides information from the development phase of the study: focus group transcripts, the survey instrument for the pilot study conducted before the main survey, and a data set containing the responses to the two dollar amounts asked about in that pilot survey.

The complete dataset obtained from administration of the main study survey instrument is contained in Appendix K (CD only). The dataset is provided in the formats of several different statistical formats: SAS, SPSS, and STATA. We hoped that by making this dataset publicly available, researchers interested in looking at a wide range of empirical issues will have a dataset from a large state-of-the-art CV study to work with.

Appendix L (book only) contains our response to the industry-sponsored critique conducted by Triangle Economic Research (Dunford *et al.*, 1996). This response discusses both general issues raised about the use of CV and specific points concerning the design of the survey instrument used in this study and the analysis of the collected data.

CHAPTER 2
Scenario Identification and Survey Design

2.1. Introduction

The COS study team undertook this research effort in order to construct a monetary measure of the total *ex ante* economic value for preventing a specified set of natural resource injuries. There are two standard (Hicksian) monetary welfare measures used by economists: minimum willingness to accept (WTA) compensation to voluntarily give up a good and maximum willingness to pay (WTP) to obtain a good. These measures are defined in relation to an economic agent, for us, the public. Which of these two is the appropriate measure depends on who holds the relevant property rights in a particular good. If the public wishes to prevent oil spills along the coast and the oil companies have a right to spill oil along the coast, the public must purchase from the oil companies their rights to spill oil; and therefore the maximum WTP of the public is the appropriate measure of how much the prevention of oil spills along the coast is worth to the public. But if the public has the right to an unoiled coastline that the oil companies must purchase in order to spill oil, the minimum WTA compensation of the public is the appropriate measure of how much the prevention of oil spills is worth to the public. Since oil companies do not have the right to spill oil along the coast and the public holds the property right to California's tidelands, submerged lands, and natural resources, WTA is the appropriate measure of economic value. However, using CV to measure WTA entails design features that are difficult to implement successfully;[1] hence, a choice measure based on WTP was adopted instead. The WTP measure used here represents a lower bound on the desired WTA measure.[2]

In CV studies, choices are posed to people in a survey; analysts then use the responses to construct monetary measures of value. Two interrelated decisions must be made in the course of designing the survey questionnaire: how to characterize the object of the choice and how to structure the context in which the choice is presented. The object of choice in CV studies consists of a change in the level of provision of a good, such as water quality. The context of the choice is the particular sequence of words and illustrations used to convey the essential information about the choice. The central part of this context, referred

[1] *See* Mitchell and Carson (1989) for a comprehensive discussion of this issue.

[2] For a theoretical discussion of WTP as a lower bound on the desired WTA measure, *see* Hanemann, 1991, and Carson, Flores, and Hanemann, 1998.

to here as the scenario, contains information relevant to the choice respondents are asked to make. The object of choice is described in detail sufficient for respondents to understand the baseline situation and what would and would not change. Frequently a plausible program to accomplish the change is described. The respondent is told how much the program would cost his or her household, how the money would be collected, and how the money would be used to effect the change. Then the respondent is given the opportunity to choose whether to pay a specified dollar amount and obtain the change in the good or to continue with the baseline provision of the good.

In the survey instrument for this study,[3] the object of choice is characterized as the prevention of injuries to wildlife and rocky intertidal and sandy beach shorelines from oil spills along California's Central Coast over the next ten years. The context in which the choice is presented includes the cumulative harm that is expected to be caused by oil spills that affect wildlife along the Central Coast over the next decade, a plausible program which would prevent this harm, and a payment mechanism whereby taxpayers would pay a one-time California income tax surcharge to set up the program; the oil companies are to pay all of the costs associated with operating the program for the next ten years. A referendum format was used to elicit respondents' choices: respondents are asked how they would vote if an election were being held today and the program would cost their household a specified dollar amount.[4] Other questions preceding and following this choice question ask about respondent attitudes, about the familiarity of the respondent with the affected natural resources, about the respondent's understanding of the assumptions underlying the scenario, and about the personal characteristics of the respondent and the characteristics of the respondent's household. During the interview, showcards (which present information in written form), maps, and drawings are shown to respondents to reinforce the information presented orally by the interviewers.

In this chapter, we discuss the development of the main study survey instrument, focusing on the development of the scenario. Section 2.2 outlines the basic objectives that guided the development process. Section 2.3 presents the basic design features adopted at the outset of this study. Section 2.4 describes the three phases of instrument development and the resolution of several survey design issues during those three phases. In the following chapter, the main study survey instrument is described in more detail along with the rationales for key aspects of the final design.

2.2. Objectives of Survey Instrument Development

Throughout the development process we were guided by the following objectives: the final instrument should be (1) consistent with economic theory,

[3] A reproduction of the main study survey questionnaire and graphics booklet can be found in Appendix A.

[4] The referendum elicitation format was recommended by the NOAA Panel (Arrow *et al.*, 1993, p. 4608).

(2) comprehensible to respondents, (3) focused on the defined set of injuries, (4) plausible in regard to the scenario and choice mechanism, and (5) perceived by respondents as neutral.

The first objective was to obtain a measure of damages with a known relationship to the ideal measure suggested by economic theory.[5] Specifically, the survey instrument was designed to enable a monetary measure of economic value to be constructed from a well-defined choice regarding a specified set of natural resource injuries.

The second objective was to use language, concepts, and questions in the survey that respondents from all educational levels and varied life experiences would comprehend. One of the primary purposes of pretesting and piloting is to test whether or not the wording in the survey instrument meets this standard.

The third objective was to focus respondents on the described set of injuries only. This objective required carefully describing the specific set of injuries in such a way as to minimize the possibility that respondents would envision a more extensive or less extensive set of injuries. In this regard, the *ex ante* nature of the survey scenario is very important: respondents are asked to make a choice concerning a program to prevent *future* oil spills. Given the nature of oil spills and oil spill prevention programs, it is reasonable for some respondents to expect more or less injuries in the absence of the program than the injuries presented in the scenario and to expect that the prevention program may be less effective than portrayed in the scenario. Both open-ended and close-ended questions were used to monitor these divergences and to assess their impact on the results.

Our fourth objective was to design a realistic and incentive-compatible choice context, *i.e.*, a plausible scenario and choice mechanism.[6] Even if respondents understand the choice, they may not consider it plausible. As noted above, we used the referendum format to elicit the respondent's choice. A large number of other design decisions made to enhance plausibility will be noted in this and the following chapter. For example, describing the State as the survey's sponsor helped enhance the referendum's realism, particularly since a one-time increase in State income taxes was the payment vehicle used in the survey. Further, the State's intent in conducting the survey was explained in such a way that respondents would find it reasonable to be asked about how they would vote on the program.[7]

5 *See* Mitchell and Carson (1989) for an overview of the economic concepts underlying monetary measures of value.

6 Many problems with contingent valuation surveys arise because respondents are asked to make choices in implausible contexts about goods that are too vaguely defined with the result that respondents may perceive their answers as unlikely to influence either provision of or payment for the good. *See* Mitchell and Carson (1989) for a general discussion of CV survey design issues and Mitchell and Carson (1995) for an overview of current CV survey design issues. Carson, Groves and Machina (1999) discuss incentive issues related to survey design.

7 *See* section 3.3.

Perceived neutrality was the fifth goal: respondents should *not* perceive the purpose of the interview as the State's promotion of a particular choice. To this end, we took care to avoid bias in the wording and sequence of material; and we encouraged respondents to consider a number of reasons they might want to vote for the program and also a number of reasons they might want to vote against it. We used follow-up questions to monitor our success.[8]

2.3. Basic Design Features

Four basic design features were adopted at the outset of this study. The first is the use of in-person interviews.[9] In-person survey administration offers several important advantages over the standard alternatives – telephone surveys and mail surveys. The presence of an interviewer helps to maintain respondent motivation for the approximately half-hour interview that is needed to present a sufficiently detailed scenario. The interviewer is able to pace the narrative to accommodate the respondent's needs and is able to punctuate the narrative with visual aids to more effectively communicate scenario information and maintain respondent interest.

Table 2.1. Sequence of Survey Components

1.	Attitudinal questions
2.	Background information
3.	Description of the scenario, including • natural resource injuries, • the program to prevent some or all of the natural resource injuries, • how the program would be paid for, • reasons to vote for or against the program, and • the cost of the program to the respondent's household
4.	Vote question
5.	Vote-motivation questions
6.	First vote-reconsideration question
7.	Vote-assumption questions
8.	Demographic and other background questions
9.	Second vote-reconsideration question
10.	Interviewer debriefing questions

The second design feature is the use of a questionnaire framework consisting of the sequence of basic survey components shown in Table 2.1.[10] In previous studies (*e.g.*, Carson *et al.*, 1992), this framework has worked smoothly for both respondents and interviewers and has yielded reliable WTP estimates. The first component of this questionnaire framework is a series of questions that measure

8 *See* section 5.3.4.

9 This mode of survey administration was recommended by the NOAA Panel (Arrow *et al.*, 1993, p. 4608).

10 The rationale for this sequence is discussed in Chapter 3.

respondent attitudes towards a variety of government-provided public goods. Next are background information, a description of the scenario, a vote question, vote-motivation questions, a vote reconsideration question, vote assumption questions, demographic and other background questions, and a second vote reconsideration question. The final component is a series of debriefing questions that the interviewer completes after leaving the respondent.

As the third design feature, we employed an escort ship program similar to that used previously in the *Exxon Valdez* Oil Spill (EVOS) study (Carson *et al.*, 1992).[11] Respondents in the COS study also found the escort ship program a plausible way to prevent harm from oil spills in a particular location and only in that location. Respondents also believed the program would be expensive to implement, a belief that helped make credible whichever of the several tax surcharge amounts respondents received.

The fourth design feature is the referendum elicitation format: the respondent is asked whether he or she would vote for or against a program at a given tax amount. Respondents find being asked about how they would vote on such a matter plausible and are reluctant to vote to tax themselves unless they are convinced that what they would get is worth that amount. In this study, respondents are also asked to explain their vote and to answer questions about their perceptions of various aspects of the scenario.[12] Respondents are also given the opportunity to reconsider their initial vote.

In addition to these four basic design features, throughout the survey development process we used a conservative design strategy: wherever the relevant facts, theory, or methodological considerations did not dictate that there was one correct design decision, we adopted the alternative that would tend to *reduce* the likelihood of a vote *for* the program and that would therefore reduce the estimated value of the program.[13]

2.4. Survey Development Work

Most of the study team for this study were also members of the study team that conducted the study (Carson *et al.*, 1992; Carson *et al.*, 2003) assessing the damages from the *Exxon Valdez* oil spill (EVOS). Our work on the EVOS study provided us with a wealth of useful information about how people perceived oil spills. The first pilot survey for the EVOS study was conducted in San Jose California, giving us first hand experience valuing oil spill prevention with a sample of California residents. Our experience on the EVOS study

[11] The escort ship program described in the *Exxon Valdez* survey was in fact later set up in Prince William Sound, the site of the *Exxon Valdez* spill, and subsequently prevented a supertanker, which had lost power, from drifting into the rocks; see, *e.g.*, L.A. Times, March 26, 1993.

[12] Vote motivation questions and scenario debriefing questions are both recommended by the NOAA Panel (Arrow *et al.*, 1993, pp. 4608–4609).

[13] This strategy was also one of the NOAA Panel's recommendations (Arrow *et al.*, 1993, p. 4608).

allowed us to start the development work for this study at a much more advanced stage. In contrast to that of the EVOS study, our objective in this study was to design a survey instrument for one area that could later, with only a straightforward set of modifications, be used to value spill prevention in another area. Thus, in contrast to most CV studies, our intent was to develop a survey instrument that was to a substantial degree portable for use in other studies valuing oil spill prevention efforts. For a variety of reasons we picked the California Central Coast area as the setting for the scenario in the final survey instrument.

The first stage of development consisted of exploratory work, primarily focus groups, to discern people's attitudes about oil spills in general, their beliefs about what specific effects oil spills have on the environment, their perceptions about oil spills in particular coastal areas, and their reactions to different scenario features. On the basis of early focus groups and the basic set of design features summarized above, we drafted a working survey instrument.

In the next stage of in-depth pretest interviews, the initial working draft of the survey instrument was continually revised. Our aims during this second stage were to confirm our focus group findings and to see whether the verbal presentation flowed smoothly, whether respondents understood the wording and visual aids, and whether respondents regarded the choice they were asked to make as a credible one.

During the third and final development stage, we conducted a series of formal pretests and a pilot study. The formal pretests looked at different geographic areas and the possibility of valuing multiple scenarios in a single survey. The pilot study instrument included only the Central Coast area. That study used more formal sampling techniques and a larger sample than the pretests. The larger number of interviewers and the longer field period made it possible to reach a more diverse sample. As a result, the pilot study provided a detailed basis for evaluating how well the survey instrument was working in the field.

Peer reviewers in resource economics, psychology, and survey research reviewed the working survey instruments at each development stage.

2.4.1. *Phase I – Focus Groups*

The design work for this study began with a series of five focus groups conducted in different locations throughout California.[14] Focus groups usually have eight to twelve participants who are led in discussion by a moderator for about two hours. The give and take in focus group discussions is an efficient way to explore what people know and think about a topic and how they might

[14] These focus groups were held in San Diego, Walnut Creek, Riverside, Sacramento, and Irvine. Later, two additional focus groups were held in San Diego and San Mateo. Focus group transcripts are provided in Appendix J1.

react to an interview about it. The people who participate are not usually a random selection from the general public, but they may be selected to represent a range of demographic categories including age, sex, and education. Insights discovered in one group can be checked in subsequent groups; and design decisions based on group discussions can be tested in pretest and pilot studies.

For this study, participants were called randomly from the telephone directory and recruited to represent a range of demographic categories, including age and education levels. Participants came to a local focus group facility for a group discussion that lasted about two hours. To avoid self-selection bias (*i.e.*, people choosing to attend or not to attend based on their level of interest in the topic), respondents were told only that the group was being conducted to gather opinions on a current state public policy issue. Following standard focus group practice, each participant was offered an incentive payment for attending, the amount of this payment depending on where the group was held and the time and day of the week. Acting as moderator, a member of the research team introduced the topics and guided the discussion.

In the first focus group, we explored participants' assumptions, knowledge, and attitudes about a number of topics, including whether they were aware of past California spills and their effects, whether they connected the extent and type of harm with the spill's location, whether they were more concerned about some types of spills than others, and whether they felt it was credible for the State to initiate a program to prevent the harm from spills in different coastal locations. In the course of the discussion, the participants raised a number of questions:

- How can scientists predict the number of future spills and where they will occur?
- Why wouldn't the program be statewide?
- How would the program be paid for?
- Why should citizens (rather than oil companies) pay to prevent oil spills?
- Would the money collected actually be used for the stated purpose?

In the next four focus groups, we checked whether the concerns raised in the first group were representative; and we explored different scenario features (*e.g.*, preventing different numbers of future oil spills and targeting specific coastal areas and types of shoreline). Beginning with the second group, we tested draft versions of the scenario description by having the group either observe or take part in a simulated interview. During each simulated interview, group members were encouraged to comment and ask questions. We increased the amount of time devoted to the interview simulations in the third and fourth groups; and the fifth group was entirely given over to this activity. The simulated interview technique was very helpful for discovering potential problems in the content and wording of the draft scenarios. By the fifth group, we had developed a complete working draft.

Several findings emerged from the focus groups. Most participants were familiar with oil spills and interested in learning more about them. Preventing future damage from oil spills along the California coast was regarded as a plausible program that participants felt comfortable favoring or opposing; and they were reluctant to spend money on this type of program unless they were convinced that the benefits would be worthwhile. Focus group participants, not surprisingly, also seemed to have preferences for spending money on programs that would prevent spills in their own geographic area. For example, participants from the San Francisco Bay area were more concerned about oil spills in the San Francisco Bay; whereas participants from Southern California were more concerned about oil spills along Southern California's coastline.

We also learned a great deal during this part of our research about how to make the prevention program as credible as possible. The group participants raised helpful questions such as who would operate the escort ships, whether the escort ships would be needed if tankers had double-hulls, and how escort ships would prevent harm once an oil spill occurred. We used this information throughout survey development to increase the program's credibility.

The first stage of our development work helped us to assess what information was important to present during the interview and which potential sources of misunderstanding necessitated special handling in the survey instrument. On the basis of this information and the basic set of design features summarized above, the initial working draft of the survey instrument was developed.

2.4.2. *Survey Design Issues*

In the next two stages of our development work, we turned our attention to improving the draft survey instrument and resolving several design issues: the nature of the payment vehicle, the number of scenarios to include, the quantity and type of information to provide, and the types of visual aids to use.

2.4.2.1. *Payment Vehicle*

The payment vehicle is the mechanism by which the program, *i.e.*, the object of choice, will be funded. The link between the payment vehicle and the program must be plausible, and it should implicitly bring the relevant budget constraints to mind. In the focus groups we explored several possible payment vehicles, *e.g.*, a special surcharge at the gas pump and higher gas and oil prices. We ultimately settled on a one-time surcharge in California income taxes. The lump-sum nature of the one-time payment has the advantage of clearly focusing respondent attention on the total magnitude of the cost of the program. It has the drawback of imposing a much tighter budget constraint on some households than a program which could be paid for over multiple years. The income tax nature of the payment mechanism has the advantage of being unavoidable and highly visible for most respondents; but for those respondents who do not pay state income taxes, this payment mechanism may not have appropriate incentives for revealing information about their true willingness to pay for the program. The rationale underlying our choice of the income tax payment vehicle and the manner in which we dealt with respondents who do not pay state income taxes is discussed further below.

2.4.2.2. *Number of Scenarios*

The purpose of the COS CV Study was "to execute and document a contingent valuation study of natural resource damages associated with offshore, coastal, or estuarine oil spills in California".[15] In order to achieve this general goal, we first had to decide how many scenarios could be successfully incorporated into a single survey instrument without compromising the plausibility of any one scenario and – a related decision – what type of spills each scenario would describe. This decision had to take into account a budget that provided for a sample size of approximately 1,000 completed interviews for the final survey, the interviews having an average administration time of approximately 30 minutes. Furthermore, this decision could not compromise any of our five design objectives.

At the beginning of this study, we made a comprehensive review of California's past oil spills and their effects and determined that, in order for a scenario to be plausible (*e.g.*, to reflect the appropriate level of specificity), the description of any spills should include information about the following: (1) the type of shoreline affected (sandy beach, rocky shore, or saltwater marsh), (2) the geographic location of the spill (far north, San Francisco Bay, Central Coast, Greater Los Angeles, or San Diego), and (3) the amount and types of harm (primarily the harm to birds and shoreline habitat). While an instrument with multiple scenarios was explored, for the type of good to be valued in this study, the multiple scenarios made the choices appear too artificial for most respondents. While multiple instruments with different scenarios was another possible option, the results of our development work suggested that it was not feasible given our sample size.

Duc to the above considerations, we decided to include a single scenario that describes injuries to a type of shoreline that is fairly representative of a vast portion of California's coastline (*i.e.*, mostly rocky shoreline with some scattered sandy beaches). This scenario was successfully tested in the pilot and used in the final survey. Some of the findings that influenced this decision are described below.

2.4.2.3. *Quantity and Type of Information*

What information is necessary for respondents to perceive the scenario as plausible? The amount of information respondents can meaningfully process is

[15] State of California, Department of Justice, Contract Number 89-2126.

limited. Presenting information beyond that threshold can lead to information overload and inattentiveness. Our goal was to include only the information that most, if not all, respondents need to make a meaningful choice. A key reason for the use of focus groups and in-depth pretest interviews during the development process is to identify this information and how best to present it. Respondents' answers to open-ended questions and their spontaneous comments help us evaluate whether relevant information is missing from the survey instrument.

In making decisions about what kinds of information to present, we were motivated by four principles. First, the scenario should describe the harm in sufficient relevant detail so that respondents understand what the program would and would not protect. For example, we describe the harm as affecting shoreline ecosystems, including saltwater plants and small animals, and as killing or injuring several types of sea and shore birds. Some focus group participants and pretest respondents thought that oil spills also harm marine mammals and fish. As any particular spill along the Central Coast may not affect these resources in large numbers, the scenario informs respondents that marine mammals and fish are usually not affected by California oil spills.

Second, the scenario should provide information about relevant substitutes and recovery times so that respondents can place the injury in context. The survey designer may have several options by which to establish the proper context. For example, depending on the available information, when evaluating the seriousness of an injury such as 12,000 bird deaths, respondents might be told about the effect such deaths would have on the species as a whole, either in terms of extinction or endangerment, or in terms of the relevant extant populations of the birds in question.

Third, other scenario elements must be described in such a way that respondents accept them as reasonable for their intended purpose. In addition to information about the good to be valued, the scenario must clearly explain how the program would accomplish its purpose.

Fourth, the material presented should be based on the best available information. The study managers provided the set of injuries to be valued in the main study survey instrument. Although endangered bird species, marine mammals, and fish have been injured in past California oil spills, the study managers excluded them to value a more typical set of injuries. We also relied on the California Office of Oil Spill Prevention and Response (OSPR) in consultation with NOAA and the United States Fish and Wildlife Service (USFWS) to identify the bird species most affected by particular types of spills and to provide us with estimates of the California populations of these bird species.

2.4.2.4. *Types of Visual Aids*

In-person interviews commonly use visual displays to provide respondents with a graphic representation of some of the material that the interviewer presents verbally. These are particularly indispensable in a survey such as this one that must describe bundles of attributes, including geographical relationships, spill effects, different species of birds, and shoreline habitats.[16] In the course of revising the instrument, we developed a set of showcards that were continually refined on the basis of respondent, interviewer, and peer-reviewer feedback. The showcards are used to display lengthy lists of closed-ended answer categories and drawings and tables that illustrate various features of the scenario. The development of the display format for the key scenario information (*i.e.*, depiction of affected shoreline habitats and bird species as well as the cumulative harm) is described below.

2.4.3. *Phase II – In-Depth Interviews and Informal Field Tests*

The goal of the second phase of our work was to take the focus group insights and the working draft and ready a working survey instrument for field testing. Through an iterative process over eight sets of in-depth pretest interviews, we revised and expanded the initial working draft. We then began testing the survey instrument in a field setting. Several pretests helped identify potential problems with the different survey design dimensions, including question wording, flow of the narrative presentation, and interviewer skip instructions.

Each set of in-depth pretest interviews consisted of five to ten interviews conducted at one or two sites by a member of the research team. The respondents were recruited by market research firms who paid the respondents to come to their facility for this purpose. This type of interview gave us the opportunity to observe the overall flow of the interview and the way respondents reacted to its various parts. Each respondent was debriefed at the end of the interview. After each interview and before conducting the next interview in the set, we revised the instrument to address any problems with wording and flow. After the set was complete, we reviewed the results for that set of interviews; and the instrument was revised further.

In the first field test, NRDA staff administered the survey to a convenience sample of 11 respondents in three San Diego neighborhoods. Based on the interviewers' debriefing comments and our analysis of the results, we made a number of changes to improve the communication of information (both verbally and visually) and the plausibility of the scenario and of the choice.

In our second field test, again with a convenience sample, NRDA staff interviewed 14 respondents. The interviewers reported that the revisions to the survey instrument improved the communication of information and ease of administration. All of the respondents seemed to understand the material; most of the respondents were either extremely or very attentive throughout the interview; and all of the respondents either gave extremely or very serious

[16] The use of extensive visual aids has a long history in CV surveys valuing changes in air and water pollution (*see, e.g.*, Randall, Ives, and Eastman, 1974).

consideration to the vote questions. However, three respondents found the survey to be too long.[17] The interviewers pointed out several places in the narrative that seemed wordy or that had lost the respondent's attention. The interviewers also mentioned that some respondents were skeptical (a recurring theme in the focus groups) about some of the information presented to them; they were particularly skeptical about the certainty with which the scenario predicted the number of future spills (with and without the program) and the predicted injuries.

2.4.4. *Phase III – Formal Pretests and the Pilot Study*

In light of our in-depth development work, we further revised the instrument prior to formal pretesting with the objective of determining the structure of the survey instrument to be used in the pilot study.[18] These formal pretests and the pilot study were designed to simulate more of the procedures that would be used in the main survey by interviewing a larger and more diverse set of respondents. Professional Westat interviewers[19] conducted the interviews for the pretests and the pilot study in primary sampling units (PSU's)[20] selected to represent the California population. Respondents meeting specified criteria were chosen at the household level on the basis of information collected with the Westat screener interview. Table 2.2 describes the basic features of the two large pretests and the pilot study. Formal pretesting was conducted in two stages (referred to as pretest A and pretest B) to give greater flexibility in testing different wording and design features. In addition to the changes noted in the table, after each field effort, we made extensive wording changes based on interviewer comments and our review of the results. We now turn to a discussion of the design features tested in each of these development studies as well as the design points used and the substantive revisions made between them.

2.4.4.1. *Design Points*

In each of these development studies, respondents were randomly assigned to two equivalent subsamples that received questionnaires differing only in the dollar amount (referred to here as design point) they were told the program would cost their household. The results for the four design points ($10, $15,

[17] The length of the survey ranged from 27 to 50 minutes with an average of 38 minutes.

[18] We occasionally conducted a few in-depth pretest interviews during this research phase to help further assess proposed changes.

[19] Westat, Inc. is one of the country's leading survey research firms and is often retained by government agencies to conduct major surveys (*e.g.*, National Medical Expenditure Survey, National Household Education Survey).

[20] The PSU's were Los Angeles City, Los Angeles County, Alameda/Contra Costa/Marin/San Francisco/San Mateo, San Diego, Sacramento/Placer/El Dorado/Yolo, Sonoma, and Del Norte/Humboldt.

Table 2.2. Phase III – Pretest and Pilot Studies

Study	Sample Size	Design Points	Scenario Design Features
Pretest A	56	\$15, \$60	The first scenario described expected harm from oil spills in Greater Los Angeles area and program to prevent this harm and was followed by a vote question. Second scenario described expected harm from spills along Central Coast and was followed by a preference question.
Pretest B	112	\$15, \$60	The first scenario was the same as Pretest A. Depending upon treatment administered, the second scenario described expected harm from oil spills along Central Coast *or* the San Francisco Bay area and was followed by a preference question if voted *for* in first scenario or by a second vote question if voted *not-for*.
Pilot	154	\$10, \$120	Single scenario described expected harm from oil spills along the Central Coast and the program to prevent this harm.

\$60, and \$120) used in the development work guided the selection of the design points for the main survey.

2.4.4.2. *Scenarios and Revisions*

One of the purposes of the two pretests was to further evaluate the feasibility of valuing two scenarios: more specifically, whether respondents would be able to put the first choice out of their minds and treat the second choice as totally new, whether sufficiently detailed scenarios could be presented within the necessary time frame, and whether asking about a second scenario would compromise in any way the credibility of the choices. A payment vehicle – a one-time increase in California income taxes – and a single show card that consolidated injury information previously shown on three different cards was also tested.

The first scenario in the Pretest A survey instrument described a program that would prevent the expected harm to sandy beach habitat in the Greater Los Angeles area from large oil spills. Respondents were then asked how they would vote if the program cost their household a specified, one-time increase in California income taxes.[21] Because much of the material presented in the first scenario was relevant to both scenarios (*e.g.*, the description of the program), the focus in the second scenario was primarily on the information that was unique to it. The second scenario described a program which would prevent the expected harm from large oil spills along the Central coast, one of

[21] Prior to the presentation of this second scenario, a series of questions about respondent assumptions regarding various aspects of the first scenario was administered.

the two other heavy traffic areas: the effects of a typical, large spill on a rocky shore habitat was described as well as the cumulative effects from three spills expected to occur in this area over the next ten years. The respondent is then asked the following preference question:

> If the cost to your household was the same, which program would you want the State to set up – the program to prevent large spills on rocky shorelines along the Central Coast or the program to prevent large spills on sandy beaches in the Greater Los Angeles area?

In Pretest B, respondents were randomly assigned to one of two treatments that differed only in the second scenario. The first scenario in both treatments was similar to that in Pretest A, *i.e.*, the object of choice was characterized as a program that would prevent the expected harm from oil spills in the Greater Los Angeles area. The second scenario in treatment 1 described a program to prevent the expected harm from oil spills along the Central Coast; and the second scenario in treatment 2 described a program to prevent the expected harm from oil spills in the San Francisco Bay area. Respondents who voted in favor of the program in the first scenario were asked which of the two programs they preferred at the same cost.[22]

In both Pretests A and B, respondent reactions to the choice question following the second scenario suggested they found the second scenario less plausible and more artificial (*e.g.* respondents raised questions about whether the state government had carefully thought out how the program would be undertaken) than the first choice question. Though we had attempted to present the second scenario in an abbreviated format, the average interview time still exceeded our limit of 30 minutes. Furthermore, the results suggested that we might be trying to convey too much information too quickly.[23]

The split-sample test in Pretest B was inconclusive; but overall, Pretest B suggested that respondents seemed to value preventing spills that injured rocky shorelines along the Central Coast and saltwater marshes in the San Francisco Bay area more than they valued preventing those that injured beaches in the Greater Los Angeles area. With the results from both Pretest A and B combined, pretest respondents opted for programs which protected rocky shorelines in the Central Coast or saltwater marshes in the San Francisco Bay area by a margin of almost two to one over a program which would protect beaches in the Greater Los Angeles area. Further, as expected, respondents' preference with respect to protecting the different coastal areas were strongly influenced by where they lived. For example, in Pretests A and B, respondents living in

[22] In Pretest A, those respondents who had just voted *against* the program to protect sandy beaches in the Greater Los Angeles area appeared to have some difficulties with the preference question asked in the latter part of the interview.

[23] Almost 70 percent of those who voted against the program described in the first scenario said they did so because they had concerns about the program, the payment vehicle, or both.

the Greater Los Angeles area were significantly more likely to prefer a program which would prevent injuries from oil spills in that area ($p = 0.023$ and $p < 0.001$, respectively). For treatment 2 in Pretest B, respondents living in the San Francisco Bay area were significantly more likely to prefer a program which would prevent injuries from oil spills in that area ($p = 0.020$).

On the basis of these results and discussions with the peer reviewers, we decided to focus only on a single scenario in the pilot study. The pilot scenario described a type of shoreline that is representative of a large portion of California's coastline: an intermingling of predominantly rocky intertidal habitat with some sandy beach habitat. In a related change, in the pilot, we dropped references to a specific number of future spills and no longer characterized spills with adjectives like "large" or "medium". Previously our practice was to describe how many spills occur presently and how many of these the program would prevent; for example, the Pretest B instrument said:

> During the next ten years, however, medium size spills will continue to occur on sandy beaches about as often as they have in the past, which is about once every five years.

Although some respondents consistently questioned the credibility of this type of forecast – "How can you know that *two* spills will be prevented?" [Italics added.] – this type of language was needed to construct a meaningful scenario for preventing a specific number of spills.

Shifting to a single location left us free to address the respondents' skepticism about the precision of the forecast. In the pilot study, respondents were not given a number of spills but instead were told that spills occur causing harm and that "spills are expected to happen every few years along the Central Coast" and that, if the program is implemented, a specific amount of harm from these spills would be prevented. Undistracted by their skepticism regarding the prediction of the number of spills, respondents' attention was focused more firmly on the harm prevented. This approach did not seem to diminish the scenario's plausibility.

Another change in the scenario was the different handling of the possibility that respondents would mistakenly believe the program would prevent the effects of a huge spill. Originally we informed respondents that an *Exxon Valdez* size spill is unlikely to happen in California and that, in the event that it did, the program would not be able to handle a spill of this size. A review of respondents' reasons for voting *for* the program in Pretest A suggested that some respondents believed they were valuing the prevention of spills larger than the program would prevent. This belief that the prevention plan might be of some use in preventing an *Exxon Valdez* type spill was not unreasonable once the possibility of an *Exxon Valdez* type oil spill was invoked. We concluded therefore that mentioning *Exxon Valdez* type spills in the interview was fomenting the very belief we were trying to avoid. The main study B-2 responses suggest that eliminating references to *Exxon Valdez* type spills, in conjunction with other wording changes, was largely effective in avoiding this problem (see section 5.2.3.2).

Pretest A marked a shift in payment vehicle from higher prices for oil products to a one-time increase in state income taxes. Having the oil companies pay the program's operating costs helped defuse protests that the cost of preventing oil spill damage ought to be borne entirely by the oil companies. To establish the propriety of the State asking the taxpayers to assume part of the burden, pilot respondents were told that "individual oil companies cannot legally be required to pay the cost of setting up the program." This rationale also helped make credible the one-time only nature of the tax:

> ... all California households would pay a special one time tax for this purpose. ... Once the prevention program is set up, all the expenses of running the program for the next ten years would be paid by the oil companies.

Respondents' aversion to taxes and their desire to have the oil companies held completely responsible for the costs of the program tend to make this tax vehicle a conservative choice.[24]

Prior to these development studies we used multiple showcards to communicate information on a particular topic in a step-by-step process.[25] This approach has the advantage of helping to break-up long portions of text in the survey. Influenced by Edward Tufte's work on the visual display of information,[26] we consolidated the showcards in the pilot and presented all of information on the likely harm to the resources at risk on a single card.[27] Over the course of the development work, we devised a visual aid (two 11 by $8\frac{1}{2}$ inch showcards displayed side by side in a flip binder book)[28] for the main survey that consolidated injury information that previously appeared on three separate cards. The final result is a black and white drawing[29] displaying a rocky intertidal and sandy beach ecosystem with resident bird, small animal, and saltwater plant species on one side and a table summarizing the cumulative harm that the

24 Moreover, some respondents in the main study believed that the oil companies would pass their share of the costs on to consumers in the form of higher gas and oil prices; hence, these respondents are less likely to vote *for* the program for this reason (*see* section 6.4.1).

25 For example, one card would show the bird species most commonly affected by a certain type of spill. Another would give a picture of the shoreline ecosystem showing the types of shoreline habitat the spill would harm. A third card would present a table summarizing the number of birds harmed by a spill and the miles of shoreline affected.

26 *See, e.g.*, Tufte, 1983.

27 Another impetus for this presentation revision was the need to further shorten the length of the interview.

28 *See* Card D in Appendix A.

29 We used black and white rather than color because it is more conservative. The use of black and white also lowered the reproduction cost.

program would prevent over the next ten years on the other side. With this single visual aid, the respondent can see most of the important scenario information at one time. During this same time we also merged two maps into a single visual showing photographs of the three types of shoreline, the location of the shoreline types along the California coast, and the two types of tanker routes (*i.e.*, super-tanker versus barge and small tanker).[30] Changes to the visual aids required some rewording and sequence changes in the questionnaire.

While the pilot study survey instrument (Appendix J2) appeared to work reasonably well, we revised it in response to extremely helpful interviewer comments during the debriefings held after that survey. A number of changes to the instrument were made during this phase with particular attention being paid to enhancing the questionnaire's clarity, making it shorter, and improving the flow. During this development phase, in-depth pretest interviews, including some that used a cognitive interview *think-aloud* approach, were also used to test revisions.[31] The rationales for our final wording choices are discussed in the next chapter.

2.4.5. *Main Study Survey Instrument*

We completed the main survey instrument in December of 1994. After additional review by the study's peer reviewers, some minor revisions were made to the instrument and the final survey instrument was tested in several in-depth pretest interviews. We then delivered the completed survey instrument with interviewer training instructions to Westat.

[30] *See* Card B in Appendix A.

[31] In addition, respondent concerns expressed in the in-depth pretest interviews about the effect of oil spills on human health led us to add a vote reconsideration question to the pilot instrument which asked respondents how they would vote if human health was definitely not affected.

CHAPTER 3
Structure of the Main Study Survey Instrument

3.1. Introduction

This chapter describes section by section the wording, format, and sequence used in the main study survey instrument as well as the rationale underlying the key features of the final design. Unless otherwise indicated, all quoted text in this chapter is from the survey questionnaire itself and is presented in a different typeface. Any questionnaire text in uppercase is an interviewer instruction not read to the respondent. The complete survey instrument, including a copy of the graphics booklet, is provided in Appendix A.

To avoid self-selection bias from people deciding to be interviewed because of their interest in the survey's specific subject matter, prospective respondents were told that the State of California was conducting the study to "collect valuable information about how you feel the state should spend tax dollars."[1] If potential respondents asked for more information about the reasons the survey was being conducted or what the survey was about, the interviewers were instructed to use only the replies provided on a laminated Q & A card.[2] For example, if a respondent asked "Why are you doing this survey?", the interviewer was to reply "The study will provide information so State policy makers can understand how people like yourself feel the State should be spending tax dollars." If the respondent asked a question like "What is this survey about?", the interviewer was to reply "This study is about the priorities of Californians and how Californians feel the State should spend tax dollars."

3.2. Section A – Initial Questions

The main interview begins with a series of questions (A-1A to A-1F) that ask the respondent how important six state-wide issues are to him or her personally.

> A-1. Let's start by talking for a moment about some issues in California. Some may <u>not</u>[3] be important to you, others may be.

[1] *See* Westat, 1995, p. 2-23.

[2] *See* Appendix B.2.

[3] Underlining was used throughout the questionnaire to indicate to the interviewer the need to emphasize certain words to help convey the passage's meaning and to hold the respondent's interest.

SHOW CARD A[4]

First, (READ X'd ITEM). Is this issue not important at all to you personally, not too important, somewhat important, very important, or extremely important? (READ EACH ITEM, BEGINNING WITH X'd ITEM; CIRCLE ONE CODE FOR EACH; REPEAT ANSWER CATEGORIES AS NECESSARY.)[5]

The primary purpose of this series of questions and the following similar series (A-2A to A-2F described below) was to encourage respondents to think about a broad range of current policy issues as a reminder that the program described later in the interview is just one of many government-provided goods. These questions also validate the pre-interview description of the survey contained in the advance letter: "This study will collect valuable information about how you feel the State should spend tax dollars."[6] The six issues in the A-1 series that establish this context are improving education, reducing air pollution, maintaining local library services, reducing crime, protecting coastal areas from oil spills, and finding ways to reduce state taxes.

The question A-2 series draws the respondent's attention to the fact that the State already spends money on a wide variety of programs by asking how important it is to the respondent that the State continue to spend money on six current programs.

A-2. The State of California spends tax money on many programs for many different purposes. I'm going to read a list of some of these programs. For each one, I would like you to tell me how important it is to you that the State continue to spend money on it.

SHOW CARD A AGAIN

First, (READ X'd ITEM). (READ EACH ITEM, BEGINNING WITH X'd ITEM; CIRCLE ONE CODE FOR EACH; REPEAT ANSWER CATEGORIES AS NECESSARY.)

The six programs are providing job training for the unemployed, providing shelters for the homeless, protecting wildlife, providing lifeguards at state

4 These boxed instructions cue the interviewer to show a particular card or map, in this case, Card A. This card lists the five answer categories, from "not important at all" to "extremely important", for questions A-1 and A-2; *see* Appendix A.

5 Following standard survey practice, the order in which the six items were asked was randomized. The interviewer was instructed to begin with the item marked "X" to minimize response order effects. Each item thus had an approximately equal chance of being asked first.

6 *See* Appendix B-5.

beaches, providing public transportation for Los Angeles, and building new state prisons.

3.3. Section A – Description of the Scenario

The presentation of the scenario, which begins at this point, provides the circumstances of the choice which may be relevant to the decision the respondent is later asked to make, *i.e.*, to vote for or against paying for the described program. Among the material presented here is a description of the three general types of shoreline found along California's coastline, the main routes oil tankers and barges take along the coastline, the types of wildlife that have been harmed in past oil spills along the Central Coast, the wildlife that are expected to be harmed by future spills in this area, and the proposed prevention program, in particular, how the program would work and how it would be paid for.

The interviewer training for this study emphasized the importance of presenting this material in a way that would maintain respondents' interest and enhance comprehension. For example, the interviewer's manual stated:

> This study may differ from most that you have conducted because the central portion of the questionnaire is a narrative you read to the respondent. In our pretests of earlier versions of this questionnaire, we found that the text reads smoothly and that most respondents find the material very interesting. Our pretests also show that reading this type of material requires a somewhat different approach than reading regular question material.
>
> The narrative material about oil spills and the harm they cause is intended to provide respondents with important background information about the choice they will be asked to make in B-1. It is important that the respondent understand what you are reading so that he/she can take this information into account when answering the voting question.
>
> Because of the amount of material you will be reading, there is a risk that some respondents may become bored or disinterested. We have found that the show cards interest most respondents a great deal and help involve them in the interview. Another lesson from the pretests is that it helps to read the material in a manner that is conversational and interesting. To do this, you need to make use of effective "body language" and use a tone of voice and manner that is interesting.[7]

Instruction boxes placed strategically in the text of the questionnaire

[7] Westat, 1995, pp. 4-4 and 4-5.

instructed the interviewers to show the respondent visual aids (*e.g.*, maps, drawings, answer categories) at the appropriate time. These aids were designed and pretested to help respondents visualize important aspects of the scenario and to help them understand the material that was being read to them. For ease of administration, the 16 visual aids (11″ by $8\frac{1}{2}$″) were individually enclosed in clear, plastic sheet protectors and bound together in a loose-leaf binder.

After the A-2 series, the scenario description begins with an introduction of the survey's subject matter. Respondents are given a rationale for being asked if they would be willing to pay for a new program:[8]

> These are just a few of the programs the State of California currently spends tax money on. Proposals are sometimes made to the State for new programs; but the State does not want to start any new programs unless taxpayers are willing to pay the additional cost for them. One way for the State to find out about this is to give people like you information about a program so that you can make up your own mind about it. (STOP)[9] Your views are useful to State decision makers in deciding what, if anything, to do about a particular situation.

In order to avoid bias due to respondents perceiving an impression, either from the wording of the survey instrument or the interviewer's demeanor, that one response is favored over another, the respondent is then told that people responding to this type of interview have different views about the proposed program:

> In interviews of this kind, some people think that the program they are asked about is not needed; (STOP) others think that it is. We want to know what you think. (STOP)

Next, a question is asked to involve the respondent in the interview:

> A-3. Have you ever been interviewed before about whether the State should start a new program?

The final part of the introduction introduces the specific program the respondent will be asked about later in the interview. The wording implies that this type of inquiry is routine:

> In the past, people have been asked about various types of programs.

[8] The text in the questionnaire (*see* Appendix A) is presented in very short paragraphs to help interviewers keep their place. In the interest of conserving space, that convention has not been maintained here.

[9] The stop sign symbol is an instruction to the interviewer to pause before continuing.

In this interview, I am going to ask you about a program that would prevent harm from oil spills off one part of the California coast. (STOP)

This section also serves to legitimize and normalize this survey as business as usual on the part of the State. The atmosphere of normality will minimize the impression that there is anything special about this particular good or the action contemplated by the State.

Respondents are also informed that they will be asked to decide whether the program should be implemented and asked for the reasons behind that decision.

I will begin with important background information. Then I will ask you whether you think this particular program is worthwhile and why you feel the way you do. (STOP)

This paragraph is intended to encourage respondents to pay attention to the scenario and the choice.[10]

The interviewer then presents the background material which begins with a description of the three basic types of shoreline found along the California coastline:

SHOW CARD B

Along the California coast, there are three different types of shoreline. ☞[11] The areas shown here in green are mostly saltwater marshes. ☞ The areas shown in brown are mostly rocky shoreline. ☞ And, the areas in yellow are mostly sandy beaches. (STOP)

Card B consists of two 11″ by 8½″ cards presented concurrently on opposing pages. The top page of Card B shows a small color picture of each type of shoreline, and the bottom page displays a map of the State showing the approximate locations of the three different types of shoreline along the California coastline.[12] Stretches of rocky shoreline are displayed in green, saltwater marsh in brown, and sandy beaches in yellow.[13] Also depicted on the bottom of Card B are "Super-tanker routes" which run from the top of the State down to the San Francisco Bay area and the Greater Los Angeles area

[10] This technique of inducing accountability at the start an interview has been shown to promote optimal respondent effort. *See* Tetlock, 1983.

[11] The hand symbol instructed the interviewer to point to the relevant feature on the show card.

[12] Because of the size of the map and the minimum width of the color band needed for the colors to be visible to the respondent, small sections of shoreline that intermingle with another type (*e.g.*, the small stretches of beaches between rocky coves) were not displayed on the card; *see* Appendix A.

[13] Each of the shoreline pictures shown in the top of Card B has a color border which matches the color used to depict that particular type of shoreline along the coast; *see* Appendix A.

and a "Central Coast small tanker and barge route" joining the San Francisco Bay area with the Greater Los Angeles area.[14]

Next, respondents are asked if they had visited California's shoreline in the last 12 months and, if so, which of the three types had they visited:

> A-4. Have you visited any of these three types of California shoreline in the last 12 months?
>
> A-5. And, which ones are those?

Information about the tanker and barge routes shown on Card B is then provided along with a brief description of how oil spills usually occur and what happens when they occur:

> Each year, tankers and barges carrying oil make about 3,000 trips in and out of California harbors and along the Central Coast. Large oil tankers called super-tankers deliver their cargo to storage tanks and oil refineries in the San Francisco Bay and in the Greater Los Angeles area. Small tankers and barges transport various types of refined oil back and forth along the 500 miles of coastline between San Francisco and the L.A. area.
>
> Tankers and barges occasionally run into things like underwater rocks, other ships, or pipelines, and spill some of their oil into the water. Unless the spill is very small, the oil can harm wildlife. After an oil spill, the company that caused it must pay to clean up as much oil as possible from both the water and the shoreline. STOP

The focus of the interview then narrows from California's coastline to the one area of particular interest in this survey, the Central Coast. In order to legitimate the attention given to just this one area of the coast, respondents are told that measures had already been taken to set up programs in both the San Francisco Bay and the Greater Los Angeles area.[15] Very general information about how frequently spills occur in the Central Coast is provided to help define the context for the policy choice.[16]

> Over the years, the State has taken various steps to prevent harm from oil spills. Recently, steps have been taken to set up programs to prevent

14 So as to not burden the respondent with extraneous information, only the major coastal oil tanker and barge routes are included in this graphic.

15 In fact, a tug-escort program, which some pretest respondents were aware of, had been initiated in San Francisco Bay.

16 Because we wanted to focus the respondent on the cumulative harm from oil spills rather than the actual number of oil spills, we were intentionally vague about the past number of spills.

harm from spills in the San Francisco Bay and in the L.A. area. The State wants to know whether people think this would be worth doing for the Central Coast. (STOP)

As you can see here, most of the Central Coast is rocky shoreline with some scattered sandy beaches. Oil spills that harmed wildlife have happened here every few years.

The next portion of the interview describes the types of wildlife that have been harmed in past California oil spills. To enhance the credibility of this information, the respondent is told the data were provided by state and university scientists.[17] A black and white drawing (Card C) of a typical Central Coast ecosystem (*i.e.*, intertidal rocky shoreline and sandy beach habitat) is used as a conservative way to portray the wildlife affected by a spill in this area. Respondents are told that none of the wildlife shown are endangered. Furthermore, respondents are also provided with information on the size of the different bird populations in California and told that these birds also live in other States.

About mid-way through this description, the respondent is asked whether he or she is familiar with any of the five bird species (that are presented as those harmed the most by past spills) shown on the card. Question A-7 and follow-up question A-8 are asked at this particular point to enhance respondent's attention to the material.

State and university scientists were asked to provide information about the effects of these past spills.

SHOW CARD C

This drawing shows the types of wildlife that Central Coast spills have harmed. It shows five types of birds and other types of small animals that live in or near the water. Take your time to look it over.

(STOP) UNTIL R IS FINISHED LOOKING AT CARD

The five birds shown here are the types of birds that past spills have harmed the most.

A-7. Do you happen to be familiar with any of these birds?

[17] As noted in Chapter 2, the set of injuries was provided to the study team by OSPR scientists in consultation with NOAA and USFWS. The survey's description of the injury and recovery time information was reviewed and approved by study managers prior to the fielding of the main survey.

A-8. Which ones? (**CIRCLE THOSE MENTIONED**)

According to scientists, none of these birds are in any danger of becoming extinct. The number next to each bird shows how many of them live in California.[18] For example, there are about 290,000 Pacific Loons and 130,000 Western Gulls.[19] All five types of birds also live in other States.[20] STOP

Whenever oil washes up on the shoreline along the Central Coast, it harms many small animals and saltwater plants. Some are shown here. They include clams, sea stars, crabs, mussels, kelp, and other seaweed. None of these are in any danger of becoming extinct. STOP

To dispel possible misconceptions about what would be prevented, information is also provided about what resources are not consistently affected by a coastal spill:

Marine mammals – such as whales, seals, and dolphins – are not usually affected by the oil because they generally leave the area when a spill occurs.[21] Fish also leave the area and are not affected. STOP

Next, the interviewer describes a context that pretesting suggested was plausible for the prevention program:

Recently, the federal government passed a new law to help reduce the number of oil spills. Ten years from now, all oil tankers and barges will be required to have two outer hulls instead of the single-hull most of them have now. Double-hulls provide much more protection against oil leaking after an accident. However, it will take ten years before all single-hulled tankers and barges can be replaced. Until then, spills are expected

[18] The number shown was either the breeding population or the peak migratory population, whichever was relevant given the species. OSPR, in consultation with NOAA and USFWS, provided the population figures as well as the five particular species that were described.

[19] Box 1 provides the interviewer with the following scripted response if the respondent asked if Western Gulls are the same as sea gulls: "Western gulls are one of a dozen types of sea gulls." The other three species shown on the Card are the Rhinoceros Auklet, Common Murre, and Brandt's Cormorant. *See* Appendix A.

[20] This statement reinforces the notion that these animals are not in danger of extinction.

[21] Box 2 provided the interviewers with the following scripted response if the respondent asked about what happens to sea otters or mentioned that he or she thought sea otters were also affected in oil spills: "Like other marine mammals, sea otters usually leave the area when a spill occurs. They have not usually been affected by past Central Coast spills." Some sea otters have been injured by oil spillage; see, *e.g.*, Mercer Management Consulting (1993). The omission of sea otters from the injury scenario results in a more conservative valuation.

to happen every few years along the Central Coast, just as they have in the past, unless something is done. (STOP)

As some respondents may be aware of the coming change in tanker technology, this information provides a rationale for why the program is worthwhile during the next ten years.

Respondents are then told that, based on studies scientists have made of past spills along the Central Coast, a certain amount of harm to wildlife is expected in this area over the next ten years. Because some pretest respondents objected to an earlier version of the scenario which specified the number of spills that would be prevented on the grounds that the number of spills did not seem credible for one reason or another, the text focuses on the expected, cumulative harm over the next ten years if some prevention program is not implemented.

SHOW CARD D

This shows the total amount of harm to wildlife that state and university scientists expect will happen in the Central Coast area over the next ten years. It is based on studies scientists have made of past spills in this area. (STOP) In the next ten years: scientists expect that a total of about 12,000 birds of various types will be killed by oil spills off the Central Coast. In addition, about 1,000 more birds are expected to be injured but survive. Also, many small animals and saltwater plants are likely to be killed along a total of about ten miles of shoreline.[22] (STOP)

The harm from an oil spill is not permanent. Over time, waves and other natural processes break down the oil in the water and on the shoreline. Typically, within ten years or less after a spill, there will be as many of the affected birds as before the spill. The small animals and saltwater plants in the affected area recover somewhat faster, in about five years or less. (STOP)

Consistent with the NOAA Panel's recommendation (Arrow *et al.*, 1993, p. 4609), several different checks on respondent understanding and acceptance of the scenario are used in this survey. One such check was the following question which gives respondents the opportunity to say whether they would like to know anything more about the harm spills are expected to cause:

A-10. Is there anything more that you would like to know about the

[22] Information about the number of miles of Central Coast shoreline was provided earlier in the interview to help place the injured area in the larger context of the uninjured area.

> harm oil spills are expected to cause off the Central Coast over the next ten years?

Those who answered "yes" were asked an open-ended question:

> A-10A. What is that?[23]

This question was the first of several in the survey which required the interviewers to record verbatim the respondent's answer. The interviewers were instructed to record on the questionnaire whatever the respondent said as closely as possible, asking the respondent to pause, if necessary, so an answer or comment could be completely transcribed. The importance of accurately recording the comments in the interview, both the answers given in response to specific questions like A-10A above and spontaneous remarks made by the respondent at any other place during the interview, was emphasized in training and in the interviewer's manual (Westat, 1995, p. 4-8.)

The interviewers were also trained to use non-directive probing techniques to clarify respondents' answers to open-ended questions if the answers were vague or did not adequately answer the question.[24] Such probing is a standard survey procedure used to refocus respondent's attention on the question and get the respondent to elaborate or think about an incomplete or irrelevant answer without influencing the content of the subsequent answer.

The next portion of the interview describes the prevention program.

> If taxpayers think it is worthwhile, the State could prevent this harm by setting up a prevention program for this part of the coast. This program would be similar to those successfully used by other states, such as the State of Washington. It would last for ten years, until all tankers and barges have double-hulls. (STOP) This program would do two things. First, it would help prevent oil spills from occurring. Second, if an oil spill does occur, it would prevent the oil from spreading and causing harm. (STOP) Here is how a Central Coast program would prevent spills from occurring.

> | SHOW CARD E |
> |---|

> Oil spill prevention and response centers would be set up in three different locations along this part of the coast. Specially-designed ships, called escort ships, would be based at each center. An escort ship would travel alongside every tanker and barge as it sails along the

[23] The interviewers were trained in answering respondents' questions, both questions asked in response to specific questions (such as A-10A) and questions asked spontaneously (*see* Westat, 1995, p. 4-7).

[24] Chapter 5 of the interviewer's manual is devoted entirely to probing; *see* Westat, 1995.

Central Coast. This would help prevent spills in this area by keeping the tankers and barges from straying off-course and running into underwater rocks, other ships, or pipelines.

SHOW CARD F

If any oil were spilled, here's how the program would keep it from spreading and causing harm. The crew of the escort ship would quickly put a large floating sea fence into the water to surround the oil.[25] To keep it from spreading in rough seas, this fence would extend 6 feet above and 8 feet below the surface of the water. Then skimmers, like the one shown here, would suck the oil from the surface of the water into storage tanks on the escort ship. Other ships would be sent from the nearest prevention and response center to aid in the oil recovery and clean-up.

Card E shows the location of the three proposed oil spill prevention and response centers along the Central Coast; and Card F illustrates with a black and white drawing how the escort ship would keep the oil from spreading and causing harm if an oil spill were to occur.[26] A two-part, open-ended question is asked next to ascertain what additional information the respondent might find relevant about how the program would work:

A-12. Is there anything more that you would like to know about how this prevention program would work?

A-12A. What is that?[27]

The payment vehicle used in this study is the California income tax. The payment is described as a one-time payment that would be in addition to what the respondent would normally pay in state income taxes.[28] The one-time household payment emphasizes the respondent's monetary obligation and is conservative relative to any payment plan that would allow the household to

[25] If the respondent asked whether a sea fence was the same thing as a boom, the interviewer was instructed to answer "yes" (Westat, 1995, p. 4-35).

[26] *See* Appendix A.

[27] In case a respondent asked about what happened to the oil, interviewers were provided with the following scripted response: "Within hours, an emergency rescue tanker would come to the scene and take the oil to storage tanks on shore" (*see* Westat, 1995, p. 4-35). In addition, if a respondent asked about how the program would be paid for or about the program's cost, the interviewer was instructed to check Box 3 and say "I will come to that in just a moment" (*see* Appendix A).

[28] If a respondent asked whether this additional tax would be withheld from his or her paycheck, the interviewers were instructed to say "yes" (*see* Westat, 1995, p. 4-43).

pay over the course of several years. Further, to avoid protest votes by respondents who felt that individual oil companies should pay for all of the program's costs, a feeling that was expressed frequently in our early pretesting, respondents are told that the oil companies could not legally be required to pay for setting up the program but could be required to pay for all the expenses of running the program once it was set up.[29] Used in our later pretesting, this cost-sharing approach was perceived as fair and seemed to reduce protest responses.

> The money to pay for this program would come from both the tax-payers and the oil companies. Because individual oil companies cannot legally be required to pay the cost of setting up the program, all California households would pay a special one time tax for this purpose. This tax money would pay for providing the escort ships and setting up the three oil spill prevention and response centers along the Central Coast.
>
> Once the prevention program is set up, all the expenses of running the program for the next ten years would be paid by the oil companies. This money would come from a special fee the oil companies would be required to pay each time their tankers and barges were escorted along the Central Coast. Once the federal law goes into effect ten years from now, all tankers and barges will have double-hulls and this program would be closed down. (STOP)

The respondent is then told we want to know how he or she would vote if the program were on the ballot in a California referendum:

> We are interviewing people to ask how they would vote on this Central Coast prevention program if it were put on the ballot in a California election. (STOP)

The referendum format is the elicitation framework recommended by the NOAA Panel (Arrow *et al.*, 1993, p. 4608).

To provide a balanced view of the choice and to focus the respondent's

[29] If, after hearing the description of the payment vehicle, the respondent expressed the view that the oil companies should pay all costs, the interviewers were instructed to check the appropriate box in Box 4 and give the following response:

> The State cannot legally force individual oil companies to pay for setting up the program. However, the oil companies can be required to pay a special fee each time one of their ships is escorted along the Central Coast. These fees will pay to keep the program operating over the next ten years.

If the respondent asked about program costs, the interviewers were instructed to check the appropriate box in Box 4 and say "I will come to that in just a moment"; *see* Appendix A.

attention squarely on the choice after the necessarily long and detailed description of the program, possible reasons to vote for and against the program are displayed on Card G and read aloud by the interviewer:

> There are reasons why you might vote for setting up this program and reasons why you might vote against it.

> SHOW CARD G

The reason offered to vote for the program summarizes the change in harm to the Central Coast that the respondent would receive for paying the specified amount of additional tax:

> The program would prevent harm from oil spills in the Central Coast area during the next ten years. Specifically, the program would:
>
> prevent the deaths of about 12,000 birds as well as the deaths of many small animals and saltwater plants along about 10 miles of shoreline, and prevent 1,000 more birds from being injured. (STOP)

Next, the questionnaire offers three reasons that the respondent might want to vote against the program. We drew on our experience in focus groups, other pretesting activities, and other past experience to select reasons that would be perceived as providing an acceptable justification for voting against the program. The first reason offered for voting against the program was that the number of birds and other wildlife that would be protected is small relative to their total numbers and that none of the species potentially affected by Central Coast oil spills are endangered:

> the number of birds and other wildlife it would protect is small in comparison to their total numbers, and none are endangered. (STOP)

The second reason explicitly reminds respondents that there may be other issues that are more important to them than this one:

> Your household might prefer to spend the money to solve other social or environmental problems instead. (STOP)

The third reason offered for voting against the program is that the cost of the additional tax payment may be more than the household wants to spend for what the program would accomplish:

> Or, the program might cost more than your household wants to spend for this.

REMOVE CARD G

Reasons to vote against the program were provided to respondents to make those inclined to vote against the program more comfortable with making and stating that choice.

3.4 Section B – Choice Questions

At the beginning of the next section, respondents are told how much the program would cost their household. Respondents were randomly assigned to one of five versions of the questionnaire which differed only by the tax amount – either $5, $25, $65, $120, or $220 – which their household would pay if the program were to be approved.

> If the Central Coast prevention program were put into place, it would cost your household a total of $[ONE OF FIVE TAX AMOUNTS]. You would pay this as a special one time tax added to your next year's California income tax.

During training, the interviewers were told that *household* has the same meaning as it had on the Household Screener and that if the household had more than one person who paid California income taxes, the tax amount would be shared among the taxpayers in the household. Since some pretest respondents expressed confusion about this, the interviewers were instructed to say to any respondents that asked: "Think of this amount as the total amount for your household" (Westat, 1995, p. 4-41).

The choice question, B-1, asks the respondent to make a decision about the object of choice, *i.e.*, to vote *for* or *against* the prevention program at the specified tax cost.[30] To make the decision as realistic and immediate as possible, the choice is posed in terms of an election being held *today*:

> B-1. If an election were being held today, and the total cost to your household for this program would be $[ONE OF FIVE TAX AMOUNTS], would you vote for the program or would you vote against it?[31]

30 Due primarily to time constraints and a decision to use the Turnbull lower-bound mean as the summary statistic (*see* Appendix F), we elected not to ask a follow-up choice question concerning the respondent's willingness to pay a second lower or higher tax amount for the program.

31 The interviewer's manual warned that a few respondents may look to them for cues as to how they should vote at this point and that,

> in fact, it doesn't matter at all how people vote; what does matter is that their answers represent their own best judgment about their actual willingness to pay based on the information provided to them in the interview and their preferences about how they should spend their money. This is why you should use a neutral tone and an unhurried manner. (*See* Westat, 1995, p. 4-41).

The respondent is offered two explicit options: voting *for* and voting *against*.[32] Some respondents do not respond to either of those two explicit options. In order to avoid pressuring respondents who give some other response, the interviewers were trained to accept other responses, such as "don't know" or "not sure," as valid answers for this question and to record them as "*not sure*" without additional probing (Westat, 1995, p. 4-43). If the respondent said something like "I don't vote," "I'm not registered," or "I'm not a citizen," the interviewers were instructed to say "If you did vote, would you vote for the program or against it?" (Westat, 1995, p. 4-43). The interviewers were also trained to handle any attempts by the respondent to ask them what they (the interviewer) think about the question by saying:

> We want to know what you think. Take as much time as you want to answer this question. (PAUSE). We find that some people say they would vote for, some against; which way would you vote if the program cost your household $____? (Westat, 1995, p. 4-43)

Depending on the response to question B-1, respondents are asked one of three follow-up, vote-motivation questions to capture the respondents' explanations for their votes, a procedure specifically recommended by the NOAA Panel (Arrow *et al.*, 1993, p. 4609). After the motivation question, those who voted *for* are also asked a question which allows them to reconsider their vote *for* the program. Those who do not vote *for* the program are not given an opportunity to reconsider their votes at this point and are only asked the appropriate vote-motivation question.

The vote-motivation question for respondents who voted *for* (B-2) is worded to assess as specifically as possible, without leading the respondent to give any particular answer, the reason that the respondent's household would be willing to pay the stated tax amount.

> B-2. People have different reasons for voting for the Central Coast prevention program. What would the program do that made you willing to pay for it? **(PROBE: Was there something specific that the program would do that made you wiling to pay for it?)**

If the respondent's answer was vague or non-responsive, the interviewer was directed to probe for a more specific answer. As noted earlier, in order to clarify vague answers, the interviewers were trained to use neutral and nondirective probes whenever respondents gave answers that were not responsive to

[32] This dichotomous choice – *for* or *against* a particular level of taxation – is recommended by the NOAA Panel (Arrow *et al.*, 1993, p. 4612). *See also* Chapter 4 of Carson *et al.* (1994; 1998) for a discussion of the NOAA Panel's recommendation to also include an explicit *would not vote* answer category.

the particular question or were vague or non-specific. The interviewers were provided with a card listing a set of standard probes and were trained when and how to probe.[33] In addition, in question B-2, a standard non-leading probe was furnished for vague responses.

After the motivation question (B-2), respondents who voted *for* the program were offered a chance to change their vote from *for* to *against*.[34] Question B-3 explicitly deals with the concern of some respondents that oil spills may affect human health. In addition, respondents who wanted to change their *for* votes for any other reason could also take advantage of this opportunity to reconsider.

> B-3. Occasionally, people vote for the program because they are concerned that oil spills may somehow harm human health. Suppose human health was definitely not affected and the program would only prevent harm to birds, small animals, and saltwater plants. Would you vote for or against the program if it cost your household $[B-1 TAX AMOUNT]?

Respondents who voted *against* the program at question B-1 are asked about their motivation in Question B-4:

> B-4. Did you vote against the program because it isn't worth that much money to you, or because it would be somewhat difficult for your household to pay that amount, or because of some other reason?

This way of asking about the respondent's motivation to vote *against* the program alleviates the discomfort some respondents might feel at revealing motivations they find unpleasant or too personal (*e.g.*, they couldn't afford to pay for the program). The interviewers had specific instructions to record verbatim all "other" answers to B-4. As a conservative measure, the option to reconsider was not offered at this point to those who voted against the program to avoid the possibility that they would feel pressured to vote *for* as a result; but they are given a chance to reconsider later at Question D-15.

Finally, respondents who answered "don't know" or "not sure" at B-1 were asked question B-5:

> Could you tell me why you aren't sure about how you would vote? (**PROBE**)

If the respondent's answer was vague, the interviewer was prompted to use a probe such as the following: "Can you tell what it is about the program that

[33] *See* Chapter 5 of the interviewer's training manual (Westat, 1995).

[34] This is the first of two reconsideration opportunities that are offered to every respondent who voted *for* the program at question B-1.

made you unsure" (Westat, 1995, p. 4-47)? As with respondents who voted *against*, respondents who were unsure how they would vote were not offered at this point an opportunity to reconsider to avoid the possibility that they would feel pressured to vote *for*; and they too are given a chance to reconsider later at Question D-15.

3.5. Section C – Perception of Expected Harm and Perception of the Program

The questions in Section C ask the respondents what they had in mind or had assumed about various aspects of the scenario when they voted on the program. If the respondents asked why they were being asked these types of questions, the interviewers were instructed to say the following:

> We find that some people have different ideas about this. It is important for us to know what you had in mind (Westat, 1995, p. 4-49).

Despite the difficulty some respondents have with this type of question, the answers nevertheless help assess whether scenario features are accepted by respondents when they voted.[35]

Question C-1 asks about respondents' perception of the extent of the harm from Central Coast oil spills over the next 10 years:

> Please think back to a few moments ago when I asked you whether you would vote for or against the program.

> SHOW CARD H[36]

> C-1. At that time, did you think the harm from oil spills in the Central Coast over the next ten years would be about the same as that shown here, or a lot more or a lot less?

The next question in this sequence asks how seriously respondents considered the amount of harm shown on Card H:

> SHOW CARD I[37]

[35] The answers to these questions are another type of check on respondent understanding and acceptance of the scenario (Arrow *et al.*, 1993, p. 4609). The difficulties inherent in interpreting responses to such questions are discussed below.

[36] Card H repeated the injury information shown on Card D; *see* Appendix A.

[37] Card I listed the answer categories for C-2 and was shown concurrently with Card H; *see* Appendix A.

C-2. How serious did you consider this amount of harm to be[38] ... Not serious at all, Not too serious, Somewhat serious, Very serious, or Extremely serious?

Question C-3 explores respondents' assumptions about the program's effectiveness in preventing harm from future oil spills.

SHOW CARD J[39]

C-3. Did it seem to you that the prevention program I told you about would be completely effective at preventing harm from Central Coast oil spills, mostly effective, somewhat effective, not too effective, or not effective at all?

Question C-4 explores respondents' expectations about the actual length of the period the state would impose the special tax.

C-4. When you decided how to vote, did you think your household would have to pay the special tax for the program for one year or for more than one year?

In order to learn about the way the respondent perceived the interview, the final several questions in Section C explore whether the respondent felt pressured to vote one way or the other by the interview. For those who felt they had been pushed one way or the other, two follow-up questions, C-6 and C-6A, asked which direction they felt pushed and what made them feel pushed.

C-5. Thinking about everything I have told you during this interview, overall did it try to push you to vote one way or another, or did it let you make up your own mind about which way to vote?

C-6. Which way did you think it pushed you?

C-6A. What was it that made you think that? (PROBE: "Can you be more specific about what you have in mind?" "Anything else?")

3.6. Section D – Questions on Respondent and Household Characteristics and Demographic Questions

The interview now shifts from retrospection about the choice decision to the collection of demographic and other information about the respondent and

[38] Text that follows "..." was presented as mixed case answer categories (recall that interviewers were instructed not to read anything that appeared in all upper case) rather than as part of the actual question text; a *NOT SURE* answer category was also included but was not read aloud; *see* Appendix A.

[39] Card J lists the answer categories for question C-3; *see* Appendix A.

the respondent's household. The first six questions in Section D ask about various types of household recreational activities.

> Now I would like to ask you a few questions about your household's recreational activities.
>
> D-1. Has anyone in your household ever driven along the Central Coast on Highway 1, the coast highway?

Those who said *yes* to D-1, were asked the next question:

> D-2. And, was this in the last five years?

All respondents were asked questions D-3 through D-6 about several other types of recreational activities.

> D-3. In the past five years, has anyone in your household gone saltwater boating or saltwater fishing?
>
> D-4. Does anyone in your household like to identify different species of birds?[40]
>
> D-5. During this past summer, about how many times did people in your household go to beaches anywhere along the California coast ... Never, Once or twice, Three to ten times, or More than ten times?
>
> SHOW CARD K[41]
>
> D-6. How often do you personally watch television programs about animals and birds in the wild ... Very often, Often, Sometimes, Rarely, or Never?[42]

The next question asks whether the respondent perceives himself or herself as an environmentalist.

> SHOW CARD L[43]

[40] If the respondent asked the interviewer what was meant by "identify different species of birds", the interviewer was instructed to provide a standard survey reply: "Whatever it means to you" (Westat, 1995, pp. 4-55 and 4-57).

[41] Card K listed the answer categories for D-6; *see* Appendix A.

[42] If the respondent asked the interviewer what was meant by "animals and birds in the wild", the interviewer was instructed to provide a standard survey reply: "Whatever it means to you" (Westat, 1995, pp. 4-55 and 4-57).

[43] Card L lists the answer categories for D-7; *see* Appendix A.

D-7. Do you think of yourself as an ... Environmental activist, a Strong environmentalist, a Somewhat strong environmentalist, a Not particularly strong environmentalist, or Not an environmentalist at all?[44]

Demographic characteristics are the subject of the next sequence of questions.

D-8. First, in total, how many years have you lived in California?

D-9. In what month and year were you born?

D-10. What is the highest year of school you completed or the highest degree you received?[45]

The next several questions deal with household finances. The interviewer first asks question D-11:

D-11. Currently, how many adults in your household, including yourself, work for pay?

The interviewer then asks the respondent to select the correct category of household income from a list of income ranges on a card.[46]

SHOW CARD M[47]

D-12. I'd like you to think about the income received last year by everyone in your household. Adding together all income for everyone in your household, which letter on this card best describes your household's total income for last year – 1994 – before taxes? Please include wages or salaries, social security or other retirement income, child support, public assistance, business income, and all other income.

Those respondents who report incomes in the lowest two categories in D-12 are asked in Question D-13 if they paid any California income taxes in 1994.

D-13. Did anyone in your household pay any California income taxes

[44] Here, as elsewhere, if the respondent asked what was meant by "environmentalist", the interviewer was instructed to respond, "whatever it means to you" (Westat, 1995, pp. 4-55 and 4-57).

[45] The interviewer coded the respondent's answer to this question into one of eleven categories ranging from "through 8th grade" to "doctorate degree"; *see* Appendix A.

[46] This is a standard survey research device.

[47] Card M lists 11 income categories ranging from "under $10,000" to "$100,000 or more"; *see* Appendix A.

for last year, 1994, by having taxes withheld from wages, retirement income, or other money received, or has anyone in your household sent, or intend to send, tax money for last year to the State with a tax form?

Question D-14 asks all respondents about their perception of their financial situation in the near future.

SHOW CARD N[48]

D-14. When you look ahead to the next few years, do you see your personal financial situation getting ... Much better, A little better, Staying about the same, Getting a little worse, or Much worse?

3.7. Section D – Reconsideration and Miscellaneous Questions

At question D-15, the respondent was offered a final opportunity to change his or her vote:

Now that we're almost at the end of the interview and you have been able to think a bit more about the situation, I'd like to give you a chance to review your answer to the voting question. You were asked if you would vote for or against a program that would prevent the harm that I showed you earlier on this card.

SHOW CARD O[49]

D-15. If an election were being held today, would you vote for the program or against the program if it cost your household a one-time tax payment of $[B-1 TAX AMOUNT]?

For those who had earlier voted *for* the program at Question B-1, Question D-15 is a second reconsideration question. For those who had voted *against* the program at B-1 and those who said at B-1 that they were not sure, D-15 is the only reconsideration question.

The last two questions in this section, D-16 and D-17, inquire which method of paying for environmental programs the respondent would prefer and, given that the State of California is identified as the survey's sponsor, how much trust the respondent places in state government.

D-16. There are different ways for people to pay for new programs to

[48] Card N lists the answer categories for D-14; *see* Appendix A.

[49] Card O repeats the same injury information shown on Cards D and H; *see* Appendix A.

> protect the environment. One way is for the government to pay the cost. This will raise everyone's taxes. (STOP) Another way is for businesses to pay the cost. This will make prices go up for everyone. (STOP) If you had to choose, would you prefer to pay for new environmental programs ... Through higher taxes, or Through higher prices?
>
> | SHOW CARD P[50] |
> |---|
>
> D-17. Generally speaking, how much confidence do you have in the California state government? Would you say ... A great deal, Some, Hardly any, or None?

Question D-17 is the last question asked of the respondent. At this point, the interviewer is asked to thank the respondent for his or her cooperation.[51]

3.8. Section E – Interviewer-Evaluation Questions

The interviewers were asked to give their impressions about certain aspects of the interview by answering the questions in Section E. All questions in this section were answered by the interviewers after they left the respondents' homes. About this section, the interviewers were told the following:

> Section E of the questionnaire is designed to provide us with your feedback. It is important that you complete this section as soon as possible after you have conducted the interview so that it is still fresh in your mind. It is crucial to the evaluation effort that you answer all applicable questions as fully as possible. We are very interested in hearing about your experiences with the materials, procedures, and questions. You, as an interviewer, are our most important source of information for evaluating how well these worked. (Westat, 1995, p. 4-69).

Questions E-1 to E-3 ask the interviewer to record (by observation) the respondent's sex and race and to transfer the respondent's zip code to the boxes provided:

[50] Card P lists the answer categories to D-17; *see* Appendix A.

[51] In those cases in which the interviewer had elected to administer the abbreviated version of the Household Screener, the interviewer is instructed to say: "I have just a few more questions I need to ask about the other adults in your household. Let me verify that there are (number from AS-1) people 18 or older living in this household." (*see* Appendices A and B.1). The interviewer then asked questions S-3 through S-5, the questions in the enumeration table, and S-12.

PLEASE NOTE THE FOLLOWING ABOUT THE RESPONDENT BY CIRCLING THE NUMBER OF THE CORRECT RESPONSE:

E-1. SEX

E-2. RACE

E-3. TRANSFER THE RESPONDENT'S ZIP CODE FROM THE ADDRESS LABEL ON THE CALL RECORD FOLDER: □□□□□

Questions E-4 to E-9A ask the interviewer to give his or her impression about how attentive to the interview the respondent had been and what difficulties the respondent appeared to have had.

E-4. What was the reaction of the respondent as you read **A-3 through A-13?** (This is the descriptive material including the maps and drawings).[52]

a. How distracted was the respondent?
b. How attentive was the respondent?
c. How interested was the respondent?

E-5. Did the respondent say anything suggesting that he or she had any difficulty understanding either the harm caused by Central Coast oil spills or the prevention program?

E-5A. Please describe the difficulties. [OPEN-ENDED]

E-6. Did the respondent have any difficulty understanding the voting question, B-1?

E-6A. Please describe the difficulties. [OPEN-ENDED]

E-7. When you asked B-1, did you feel the respondent was impatient to finish the interview?

E-7A. How impatient was the respondent?

E-8. How serious was the consideration the respondent gave to the decision about how to vote?

52 The scale includes the following categories: extremely, very, somewhat, slightly, not at all, and not sure; *see* Appendix A.

E-9. Not counting you and the respondent, was anyone age 13 or older present when the respondent voted?

E-9A. Do you think the other person(s) affected how the respondent voted or don't you know?

The final question is an open-ended invitation to the interviewer to make any other comments about the interview:

E-10. Do you have any other comments about this interview?

No specific instructions were provided other than "record any other comments you think would be useful here about the interview" (Westat, 1995, p. 4-75).

CHAPTER 4
Implementation of the Main Study Survey

4.1. Introduction

Westat's implementation of the main study survey consisted of several steps. In preparation for fielding the survey, a random sample of dwelling units (DU's) was drawn; an interviewer's training manual was prepared; and Westat's interviewers attended a two-day training session.[1] During the 14 weeks of main survey data collection, the interviewers were supervised by regional field supervisors and a project director. As interviews were completed, Westat conducted quality control edits and validation interviews. At the end of the data collection, sample weights were constructed by the Survey Research Center at the University of Maryland. Finally, data sets containing the responses to both the closed-ended and open-ended questions were prepared. This chapter provides a detailed discussion of each of these steps.

4.2. Sample Design

The multi-stage area probability sample drawn for the COS main study represents the population of English-speaking Californians, aged 18 or older, living in private residences they own or rent (or to whose rent or mortgage they contribute). A sample of dwelling units was drawn from areas randomly selected in the fall of 1991 for inclusion in the National Adult Literacy Survey (NALS).[2] There were three stages of sample selection: first, primary sampling units (PSU's) consisting of one or several counties were selected; then segments consisting of Census blocks or block groups within the PSU's; and finally dwelling units (DU's) were selected within the segments.

COS used the same sample PSU's selected for NALS.[3] Within these PSU's, 167 segments were chosen with probabilities of selection proportionate to

[1] The interviewer's training manual is provided in Appendix B.8. It contains additional details on a number of topics covered in this chapter.

[2] NALS was conducted by the Educational Testing Service and Westat for the U.S. Department of Education. We would like to thank the Adult Education Unit of the California Department of Education and the California State Library for granting Westat permission to use a random subset of the NALS sample listings that were not selected for that study.

[3] The ten PSU's were Sacramento, San Francisco/Oakland, Riverside/San Bernadino, Los Angeles City, Los Angeles County/Long Beach, Anaheim/Santa Ana, San Diego, Del Norte/Humboldt, Sonoma, and Bakersfield.

segment population. In the third and final stage of sampling, a random selection of dwelling units was drawn. The number of DU's selected (1,747) took into account expected rates of occupancy (some DU's would be vacant), eligibility (some would not contain any English-speaking adults), and nonresponse (some would not provide an interview) so as to yield approximately 1,000 interviews. The selected DU's were randomly assigned to one of the five tax amount treatments described in Chapter 3.[4] Within each DU, one respondent was chosen from the eligible members of the household using a random sampling table generated prior to the respondent selection.

Interviewing took place over a 14-week period from January 30 to May 3, 1995. At the beginning of this period, interviewers followed a standard probability procedure[5] to sample DU's not included on the original listing of DU's.[6] The procedure corrected in an unbiased manner for DU's missed by the NALS listers as well as for any units constructed after the listing was conducted. Fifty additional DU's were added to the sample as a result of implementing this quality control procedure; thus the total sample consisted of 1,797 dwelling units.

4.3. Interviewer Training

The 33 professional interviewers participating in the study attended a two-day in-person training session on January 28–29, 1995, in San Diego, California. All of the interviewers had prior household interviewing experience. The training session was conducted by the study's project director Naomi Everett, who was assisted by the two Regional Field Supervisors. Westat's vice-president of survey operations Martha Berlin also attended. The study was referred to as the State Policy Study (SPS); and the interviewers were told the study was being sponsored by various California state agencies.[7]

The interviewers had been given an initial set of study materials to read before attending training. The training consisted of scripted lectures, exercises, interactive small group sessions, and role-playing sessions (using prepared scripts) in which one trainee took the role of the interviewer and another, the role of the respondent.

After introductory remarks, the first morning of the training began with an overview of the study and the role of the interviewer. Next, a demonstration interview was conducted to model the proper administration of the main survey

4 *See* section 3.4.

5 A copy of the Missed DU Procedure form can be found in Appendix B.1.

6 The Census Bureau's definition of a dwelling unit was used: a house, an apartment, or group of rooms or a single room occupied as separate living quarters, that is, the occupants do not live and eat with any other person in the structure, and there is direct access from the outside or through a common hall or area (Westat, 1995, pp. 2-10, 2-11).

7 *See* Appendix B.2.

instrument. That was followed by a lecture on locating selected DU's and screening procedures. The rest of the first day was spent reviewing techniques for administering the main interview, practicing these techniques in an interactive session, and reviewing probing techniques.

The second day of training began with a further review of screening procedures and an interactive session on administering the screener. That was followed by a lecture on how to avoid refusals in the field, a review of the probing exercise that the interviewers were given to complete the night before, and a discussion of administrative procedures. The remainder of day two was devoted to role-playing sessions. Finally, the interviewers were told to practice administering the survey instrument at home before attempting interviews at sampled DU's.

4.4. Interviewer Supervision

All interviewers reported to one of two regional field supervisors, who in turn reported to the project director. Supervisors were responsible for conferring with the interviewers regularly, reporting on and managing progress in the field, performing quality control edits, and validating interviews.

Interviewers reported to their supervisor by telephone at least once a week. The discussions included a case-by-case review, feedback on quality and production, and strategy for the remaining assignment. In addition, interviewers participated in conference calls with other interviewers and supervisors to share strategies on obtaining respondent cooperation.

Supervisors entered data on the interviews produced, time, and expenses into a machine-readable file that was designed to generate weekly field status reports. Supervisors also reported weekly by telephone to the project director on survey progress, case assignments, and refusal conversion strategies.

4.5. Quality Control

Field interviewers sent completed interviews directly to their respective supervisor. Upon receipt, the supervisors were responsible for quality control, including an edit of a percentage of each interviewer's questionnaires for completeness and accuracy in following procedures and skip patterns.[8] The questionnaires were then sent to Westat's home office for further editing. Results of the edits were discussed with the interviewers.

The edits uncovered four cases in which the respondent selection within the household was carried out improperly. None of these cases were included in the final data set; they were counted as *other non-response* to the main interview.

[8] The form used for editing is shown in Appendix B.3.

4.6. Validation of Interviews

Validation of ten percent of each interviewer's work (with the one exception described below) was conducted either by telephone by the supervisors or, for non-response cases and for households without a telephone, by an in-person visit by another interviewer working in the same PSU. The cases to be validated were randomly pre-selected in advance of the fieldwork and were performed using the form shown in Appendix B.4.

All the selected cases were successfully validated except one. Age discrepancies were noted on one non-response case; and as a result, all of the cases of the interviewer who handled that case were reviewed. This review revealed a total of two interviews where the interviewer altered the ages of the household members listed on the screener so that the person who was home at the time and available to be interviewed would be selected to be the respondent. Since information about the interview topic was conveyed to the household in the course of confirming that a problem existed with these two cases, the possibility of self-selection bias contaminated these two cases. In order to avoid self-selection bias arising from knowledge of the survey topic, no attempt was made to interview the correct respondent in these households. Instead, the two cases were treated as *other non-response* to the main interview. The remaining cases of that interviewer's assignment were validated successfully.

4.7. Sample Completion

The household screener was designed to collect information on household composition and to select a main interview respondent randomly from the eligible members of the household.[9] The disposition of the total sample of 1,797 cases follows:

Screeners Completed	1,311
Not an Occupied Dwelling Unit	219
Language Barriers	31
Refusals	175
Physical/Mental Handicaps	12
Never Reached	39
Other Non-response[10]	7
Other Ineligibles[11]	3
TOTAL	1,797

9 A copy of the SPS Household Screener can be found in Appendix B.1. The other field materials (*e.g.*, advance letter, refusal conversion letters, "Sorry I Missed You" card, "No Hablo Español" card) used by the interviewers and, when appropriate, mailed to the selected dwelling units can be found in Appendices B.5 to B.7.

10 This category includes five cases where the household moved before the screener questionnaire could be administered; one case where the only resident of the DU was deceased; and one case where the renters were temporarily residing in the DU while their permanent home in California was undergoing repairs.

11 This category consists of two DU's that were occupied on a temporary basis by visitors who resided outside of California; and one case where the only resident of the DU was living away from home in a substance abuse treatment facility.

where the *refusal, handicap, never reached,* and *other nonresponse* categories collectively represent 233 households with unknown eligibility for the survey.

The final disposition of the 1,311 cases in which a screener was completed and a main survey respondent selected follows:

Main Interviews Completed	1,085
Refusals	114
Language Barriers	65
Physical/Mental Handicaps	9
Never Reached	28
Other Ineligibles[12]	3
Other Non-response[13]	7
TOTAL	1,311

The response rate is the number of completed main interviews divided by the number of eligible households. Since the sample was intended to represent households of English-speaking Californians, aged 18 or older, living in private residences they own or rent, the ineligible cases are not included in the response rate calculations. Computing the response rate involves making an assumption about the eligibility of the 233 occupied dwelling units that were non-responses to the Screener for other than language reasons. The standard survey practice is to assume the same proportion of these cases was eligible as for those cases whose eligibility was determined during screener administration (Council of American Survey Research Organizations, 1982), which in this instance is 92.4 percent.[14] Using this approach, the response rate is 74.4 percent: 1,085 divided by $[1{,}797 - (219 + 31 + 18 + 3 + 65 + 3)]$.[15] In calculating the response rate, we removed from the denominator all the ineligible cases: 219 addresses that were *not occupied* DU's, 31 *language barriers* on the screener, 18 cases representing our best estimate of the *ineligibles* among the screener non-responses,[16] three *other ineligibles* on the screener, 65 *language barriers* on the main interview, and three *other ineligibles* on the main interview.[17]

12 This category consists of three cases where the residences were being occupied on a temporary basis by out-of-state residents and had no permanent residents.

13 This category consists of four cases in which respondent selection within the household was carried out improperly, two failed validation cases, and one case where the selected respondent's whereabouts were unknown to the other residents at the DU.

14 Of the 1,345 occupied DU's whose status was determined (*completed screeners*, screener *language barriers*, and screener *other ineligibles*), 1,243 (or 92.42 percent) were members of the eligible population (1,345 minus the screener *language barriers*, main interview *language barriers*, screener *other ineligibles*, and main interview *other ineligibles*).

15 Response rates by PSU are provided in Appendix B.9.

16 This estimate is obtained by multiplying $(1 - 0.9242)$ by 233.

17 This calculation is equivalent to the calculation of Response Rate 3 of the AAPOR standard outcome rates for in-person household surveys (AAPOR, 2000). The lower-bound estimate of the response rate, assuming that all of the 233 unknown eligibility cases were in fact eligible, is 73.5 percent. This response rate is equivalent to Response Rate 1 of the AAPOR standard outcome rates.

4.8. Sample Weights

As information about the survey topic was not provided to individuals until after the main interview began, willingness to pay for the prevention program could not have directly affected whether or not a household cooperated. However, other characteristics, *e.g.*, household location, may have been related to status as a response or nonresponse. Thus, the composition of the interviewed sample could differ from that of the total sample initially chosen. In addition, the composition of the sample initially chosen might differ from the total population by chance. These issues were addressed through weights constructed by the University of Maryland Survey Research Center.[18]

In order to correct for nonresponse, each interview was multiplied by the ratio of the number of interviews that would have been completed in the segment had there been no nonresponse divided by the actual number of completed interviews in the segment.[19] For example, if all sampled households in suburban San Francisco segments responded to the interview but only half of those in the city of San Francisco did, each of the city cases would be weighted at 2.0 and each of the suburban cases at 1.0.

In order to correct for possible chance variations that remained after the nonresponse adjustment was applied, the weights were post-stratified with respect to type of dwelling unit (a household-level variable that was measured in both our survey and in the 1990 Census). Disproportionately more interviews were conducted in single-family detached units; therefore these interviews were weighted down whereas other types of housing structures were correspondingly weighted up.

4.9. Data Entry

After review at the home office, Westat sent the completed questionnaires to NRDA for data entry. Upon receipt at NRDA, staff logged the questionnaires and entered the numeric and open-ended responses into a machine-readable file. NRDA staff corrected skip pattern violations and recording errors. A computer program was designed to assign a value of "9" (categorized as *not ascertained* in the Appendix C tables) to those questions that the respondent was not asked but should have been asked. A value of "." (*i.e.*, missing) was assigned to those questions that the respondent was asked but should not have

[18] A memo from the University of Maryland's Survey Research Center describing the construction of the weights is provided in Appendix B.10.

[19] This ratio was capped at 2.6 to reduce the impact on the sampling error due to variance in the weights.

been asked. The cleaned data set was used in the analysis reported elsewhere in this report. Tabulations, both unweighted and weighted, of the cleaned data set are found in Appendix C while the responses to the open-ended questions may be found in Appendix E.

CHAPTER 5

Evaluation of Open-Ended, Vote Assumption, Reconsideration, and Interviewer Evaluation Questions

5.1. Introduction

This chapter examines the measures related to the reliability of the choice data. In section 5.2, responses to selected open-ended questions are examined. The primary focus is on the open-ended, follow-up questions recommended by the NOAA Panel that ask respondents to explain their reasons for voting *for* or *against* the Central Coast prevention program or for not knowing how they would vote. Section 5.2 also examines the responses to questions embedded in the presentation of the scenario. In section 5.3, responses to the vote-assumption questions are examined to explore how respondents perceived various aspects of the scenario and whether they felt pressured to vote a particular way. Section 5.4 explores the characteristics of those respondents who changed their initial vote when they were given opportunities to reconsider. In section 5.5, interviewer assessments of various aspects of the interview are examined; and finally, section 5.6 presents a summary of our qualitative analysis of reliability.

5.2. Open-Ended Questions

The several questions examined in this and the following sections gauge, for example, whether respondents paid attention and took the choice opportunity seriously, whether respondents' decisions reflect their perceptions of the object of choice and their preferences for it, and whether extraneous factors influenced respondents' choices. Concern for the meaningfulness of the respondents' voting choices underlies these questions, a concern that motivated both the NOAA Panel's methodological recommendations and the design and administration of this study.

5.2.1. *Coding of Open-Ended Questions*

Verbatim responses[1] to the open-ended questions A-10A, A-12A, B-2, B-4, and B-5 were assigned to categories devised for each question based on an examina-

[1] *See* Appendix D.

tion of the responses in the first batch of questionnaires.[2] Each separate thought was assigned to one of the categories developed for that question.[3] Because a response may consist of several different thoughts and the percentaging base used in the tables below is the number of respondents who responded to the question, the percentages may sum to more than 100.

5.2.2. *Questions During Presentation of the Scenario*

The first section of the interview contained two sets of paired questions, A-10/A-10A and A-12/A-12A. The first question of each pair asked respondents whether they would like to know anything more about the material that had just been presented. Those who said *yes* to either of the first questions were asked its respective follow-up question: "What it that?" The responses to these questions can provide insight into how respondents' reacted to the information on the expected harm and the prevention program and whether they had difficulty understanding this part of the presentation.

After the description of the expected harm, Question A-10 asked if there was "anything more that you would like to know about the harm oil spills are expected to cause off the Central Coast over the next ten years?"; 85.5 percent said *no*. The responses to the follow-up question A-10A by the 142 respondents who said they wanted more information are summarized in Table 5.1.[4] The percentaging base used in Table 5.1 is the total number of respondents rather than those who answered affirmatively the preceding filter question, A-10.

A little over 8 percent of the sample requested more detailed information on a specific aspect of the harm (*e.g.*, affected wildlife, recovery times) or questioned the possible harm to humans or the economy, including recreational use of the affected shoreline.[5] About 2.1 percent of the sample asked about the validity and source of the information presented and, related to that, the sponsor of the survey. Another 3.4 percent inquired about an aspect of the harm that had already been described or an element of the scenario that was to be presented in a later section of the questionnaire (*i.e.*, questions about cost or what could be done to prevent the harm).[6]

2 The only other open-ended question in the survey was C-6A that asked the respondent why he or she felt pushed to vote one way or another. Because only the 5.5% of the sample who felt pushed were asked this question, the responses to C-6A were not coded into formal categories. *See* section 5.3.4 below.

3 In many cases, the open-ended responses were recorded in short sentences; hence the task of identifying separate thoughts was relatively straightforward. *See* Appendix E.

4 Fifteen responses given at A-10A were not queries about more information and are excluded from Table 5.1.

5 The issue of possible effects on human health is addressed later in the main interview at question B-3; *see* section 5.4.

6 If the answer to the question had been covered in material presented earlier, the interviewers were to go back to that material and re-read the pertinent information. If the question would be covered later in the questionnaire, interviewers were instructed to say: "*I will come to that shortly*" (Westat, 1995, p. 4-7).

Table 5.1. Respondents Who Asked Questions at A-10A as a Percentage of Respondents Who Answered A-10 [N = 1,085][(a)]

WANTED MORE INFORMATION ABOUT ...		
HARM OR POSSIBLE HARM		8.30%
Some aspect of harm not described.	4.70%	
Possible impacts on humans/human health/drinking water.	2.86%	
Possible impacts on economy/recreational use/tourism.	0.74%	
LEGITIMACY		2.12%
Validity/source of information.	1.75%	
Sponsor of survey.	0.37%	
PREVIOUS/SUBSEQUENT ITEMS		3.41%
Some aspect of harm already described.	1.84%	
Cost of program/who pays for program?	0.74%	
Can anything be done to prevent harm?	0.83%	
OTHER[(b)]		1.84%
DID NOT REQUEST MORE INFORMATION		85.53%

(a) Percentaging base is 1,085 (*i.e.*, all respondents). Percentages total more than 100 percent as multiple responses were allowed.
(b) Only includes those for whom no other category was coded.

Question A-12 followed the description of the prevention program. When asked "Is there anything more you would like to know about how this prevention program would work?", 78.7 percent said *no*. The responses of the 218 respondents who wanted more information are summarized in Table 5.2.[7] As in the preceding table, the percentaging base is the total number of respondents, not merely those who answered affirmatively the preceding filter question A-12.

As one might expect, the most commonly asked question concerned the cost of the program or who would pay for it: 8.3 percent asked about this at A-12A. This information was presented in the very next section of the interview. The second type of query, asked by 8.3 percent of the sample, was either about various aspects of how the program would work (*i.e.*, a feature already described or additional information about a specific feature of the program that was not discussed) or about other alternative programs. The third type, asked by 1.7 percent, related to the program's effectiveness and often involved expressions of skepticism about whether the program would actually work. Last, 1.7 percent asked about what happens to oil after it is spilled or about what happens to oil after it is recovered by the escort ships.[8]

7 Thirteen responses given at A-12A were not queries about more information and are excluded from Table 5.2.

8 If the respondent asked about what happens to the oil, the interviewers were instructed to answer: "*Within hours, an emergency tanker would come to the scene and take the oil to storage tanks on the shore*" (Westat, 1995, p. 4-35).

Table 5.2. Respondents Who Asked Questions at A-12A as a Percentage of Respondents Who Answered A-12 [N = 1,085][a]

WANTED MORE INFORMATION ABOUT ...		
COST OF PROGRAM/WHO PAYS?		8.29%
PREVENTION PROGRAM		8.29%
Additional information about specific feature of program (*e.g.*, escort ship, sea fence).	5.16%	
Feature of program already described.	1.57%	
Possible alternative solutions/programs.	1.57%	
PROGRAM EFFECTIVENESS		1.66%
WHAT HAPPENS TO SPILLED/RECOVERED OIL?		1.66%
OTHER[b]		2.67%
DID NOT REQUEST MORE INFORMATION		78.71%

[a] Percentaging base is 1,085 (*i.e.*, the number of respondents who answered the preceding filter question, A-12). Percentages total more than 100 percent as multiple responses were allowed.
[b] Only includes those for whom no other category was coded.

5.2.3. *Vote-Motivation Questions*

Voting choices should usually relate to (1) what respondents perceive the program to offer, (2) the cost of the program to the respondent's household, and (3) the respondent's preferences for environmental amenities of this sort.[9] Particularly relevant to the first two items are the respondents' answers to the open-ended introspection questions immediately following the choice questions. These follow-up questions in Section B asked respondents to explain why they voted as they did.[10] B-2 was asked of those who said they would vote for the program; B-4 was asked of those who said they would not vote for the program; and B-5 of those who said they were not sure how they would vote. Before examining the responses to these introspective questions, the nature of such responses is examined.

5.2.3.1. *Interpreting Introspective Questions*

Introspective questions can provide useful information; however, such questions and their responses have inherent limitations. For reasons partially discussed below, the recorded responses to such questions need not be a complete and fully accurate accounting of all factors that shaped people's judgments. This limitation imposes practical constraints on some uses of such questions.

9 Important sources of evidence for these relationships are presented in Chapter 6, including the construct validity equation and the sensitivity of respondents' choices to the dollar amounts they would pay.

10 The NOAA Panel recommended the use of such questions and the careful coding of responses to show the types of responses (Arrow *et al.*, 1993, p. 4609).

First, a number of psychological studies indicate that although people generally have good insights into their likes and dislikes and can report their attitudes well, the process *underlying* their thinking is more difficult to elicit (Nisbett and Wilson, 1977). Further, studies suggest that people are sometimes unaware of factors that shape their judgments (*e.g.*, Nisbett and Wilson, 1977) or sometimes forget what factors influenced judgments made previously (Lodge, McGraw, and Stroh, 1989). Therefore, when asked why they voted as they did in this survey, some respondents may fail to mention considerations that shaped their voting decisions and may mention factors that were not significant causes.

Second, in typical everyday conversations, speakers tend to conform to certain conversational norms or conventions (Grice, 1975). These conventions seem to affect respondent answers in survey questionnaires. For example, one such conversational convention is the implicit understanding that one should not waste time saying things that one's conversational partner already knows.[11] In this survey, when explaining their decisions to vote in favor of the program, respondents sometimes made general statements such as "the program will help the environment" which were probably intended to be understood in the context of the information that was shared with the interviewer. Because the interviewer would have just finished presenting the details of the scenario to the respondent, a respondent would presume that the interviewer was already aware of all those details. Outside of a testing situation, most respondents would have considered it inappropriate to repeat to the interviewer all the details of the scenario.

To explore whether this convention of conversational parsimony affects responses to our introspective questions, we tested whether respondents would explain the reasons for their vote in greater detail if they believed their conversational partner, the interviewer, did not necessarily share all their knowledge about the program. One way to test for such an effect is to have a different interviewer ask the *vote-motivation* questions, rather than the interviewer who had originally explained the prevention program to the respondent.

We retained a marketing research facility to recruit adult respondents (from their database of San Diego area residents) to be interviewed at their facility. The respondents were not told the specific topic of the interview in advance, only that the discussion would focus on a current state issue. During these sessions, respondents were randomly assigned to one of two conditions: (1) a slightly reduced version of the main study questionnaire[12] administered by a single interviewer (the *control* condition) or (2) this same version administered

[11] Grice provides for this convention in his discussion of two maxims: "[d]o not make your contribution [to the conversation] more informative than is required" and "[b]e relevant".

[12] This version used the same questionnaire up to and including the vote question and vote motivation questions as that used in the main study. To shorten the interview session, some of the questions asked after the vote-motivation questions in the main study questionnaire were not asked, including the reconsideration questions and household recreational questions.

by two interviewers (the *treatment* condition). At the beginning of a treatment interview, the respondent was told:

> This session will be in two parts. First, I'll give you some information about a specific situation. When I've finished, someone else will come in to ask you a few additional questions.[13]

Then the first interviewer administered the questionnaire up to and including B-1, the vote question (the amount of the additional tax payment was $25 for all respondents). After the first interviewer left the room, a second interviewer immediately came in to complete the questionnaire, beginning with the appropriate *vote-motivation* question.

The interviewers were all experienced in-person interviewers familiar with administering the main study instrument. They were randomly assigned to roles in the individual sessions, and none were informed about the particular hypothesis being tested. The interviews were conducted in a research facility that permitted continuous quality monitoring of the data collection process via one-way mirrors and sound systems.

As expected, given that the control and treatment interviews were administered identically up to and including the vote question, there was no significant difference between the percentage voting *for* the program in the two conditions ($t = 0.86$; $p = 0.39$). Sixty-nine percent of 94 respondents in the control condition voted *for* the program, compared to 61 percent of the 84 respondents in the treatment condition.[14]

Our analysis focused on respondents voting *for* the program. In both conditions, respondents voting *for* the program were asked first:

> B-2. People have different reasons for voting for the program that was described to you. What would the program do that made you willing to pay for it?

After respondents appeared to be completely finished answering B-2, in both conditions every respondent was asked:

> B-2A. Could you be any more specific about what the program would accomplish that made you decide to vote for it?

To obtain an accurate record of everything the respondent said to these two

[13] In order to keep the treatment protocol as close to the control as possible for purposes of comparability, we did not offer any additional explanation about why two interviewers were involved or about how knowledgeable the second interviewer was about the program.

[14] Furthermore, the basic demographics of the control and treatment conditions were statistically equivalent: gender ($t = -0.53$; $p = 0.60$), education ($t = 0.49$; $p = 0.62$), age ($t = -0.49$; $p = 0.62$), and income ($t = 1.10$; $p = 0.27$).

open-ended questions, the interviews were tape-recorded; and the answers to questions B-2 and B-2A were transcribed.

The simplest approach to testing our hypothesis is to examine the numbers of words respondents used to explain their voting decisions. If changing the interviewers led respondents to believe that the interviewer asking the *explanation* questions might not share all of their knowledge about the prevention program, then respondents in the treatment condition should have provided more extensive explanations. While one might argue that the treatment respondents might merely make a different explanation using the same number of words, we thought it likely they might need to use more words as well. And indeed, respondents in the treatment condition used 33 percent more words than the control group. Respondents in the control condition explained their votes with an average of 149 words ($n = 61$), whereas the treatment condition respondents did so using an average of 198 words ($n = 50$). These means are significantly different from one another: $t = 2.35$, which has a p-value of 0.021 for a two-sided test and a p-value of 0.010 for the one-sided test suggested by the conversational parsimony hypothesis.[15]

This result supports the notion that respondents in CV studies such as this one are less complete in explaining their vote decisions to the interviewer than they would be to someone else because of the conversational context in which those explanations are solicited.[16] Because respondents believe they share information about the prevention program with the interviewer, they are likely to feel it is inappropriate or unnecessary to provide as much information in explaining their vote decisions to the interviewer as they would supply to another person.

Because of the constraints on their interpretation, particularly that from introspective inaccessibility and that of conversational conventions, answers to the vote-motivation questions in our main study should be viewed as providing insight into, though not necessarily a complete accounting of, the factors influencing respondents' choices. That is, considerations mentioned by respondents should be interpreted as *traces* of the judgmental process underlying their choices rather than as comprehensive accounts of that process. The interpretation and use of respondents' responses to introspective questions such as B-2 must take their limitations into account. Despite their constraints, such questions and their answers are essential to survey design; and, in the aggregate, they are indicative of the overall validity of the study.

5.2.3.2. *Reasons for Choosing to Vote* For *the Program*

Question B-2 was asked only of those respondents who voted *for* the program at B-1:

[15] A Wilcoxon test on the ranks of the word counts also rejects the equivalence of the two distributions with a z-test statistic of 2.18 ($p = 0.029$).

[16] The result in this experiment was obtained despite the possibility that respondents may have assumed that the second interviewer was also familiar with the information presented.

Table 5.3. Reasons for Choosing to Vote *For* the Program

B-2. What would the program do that made you willing to pay for it?

CODING CATEGORY	PERCENTAGE [N = 552][(a)]
Protect the wildlife and/or affected environment.	76.09%
Cost affordable/reasonable.	17.75%
Respondent personally concerned about wildlife/environment or perceives household would benefit in some way.	14.31%
Program would work.	10.69%
Others such as grandchildren or people living in the area would benefit.	9.42%
Prevent possible physical harm to respondent or others.	7.97%
Feel responsible to help prevent harm.	6.16%
Would make oil companies more responsible.	4.53%
Might help other specific animals not mentioned in survey.	4.35%
Prevent possible permanent harm.	2.54%
Protect environment in general.[(b)]	1.99%
Other.[(c)]	2.17%

(a) Percentaging base is the number of respondents who gave a response to this question. Percentages total more than 100 percent as multiple responses were allowed.
(b) Only includes those for whom no other response category was coded or for whom the only other response was coded "other".
(c) Only includes those for whom no other response category was coded.

> People have different reasons for voting for the Central Coast program. What would the program do that made you willing to pay for it?

This particular wording was designed, in light of the probability that respondents would follow the convention of conversational parsimony and not restate to interviewers the details of the program. to focus the respondent on the outcome of the program. The interviewers were trained to use neutral and nondirective probes when respondents gave answers that seemed vague or nonresponsive to the question to determine whether the respondent had anything more specific in mind.[17]

Each distinct idea in the responses to B-2 was coded into the categories listed in Table 5.3. The percentage distribution across the categories for the 552 respondents who answered this question shows that a large majority, 76.1 percent, said they voted for the program to protect the wildlife or environment from the oil spills that the program would prevent along the Central Coast over the next ten years. About 18 percent responded that they felt the cost of the program was reasonable given what the program would accomplish. This response was significantly more likely to be given by respondents voting for the program at the two lowest tax amounts, $5 or $25, compared to those voting for the program who

[17] An example probe was included in the questionnaire after B-2: **"Was there something specific that the program would do that made you willing to pay for it?"**

received one of the other tax amounts ($p = 0.035$). The third most common category (14%) of respondents expressed personal interest in the program because it would accomplish goals that were important to the respondent. These reasons were commonly preceded by "I" or "we" (*e.g.*, "I'm a bird lover and I don't want that harm to birds" or "we are always going to the coast").[18] Other reasons in this category reflected the respondents' personal interest in water-based recreation activities such as fishing or swimming.

The reasons coded in the *program would work* category (10.7%) involved something intrinsic to the program itself, *e.g.*, the response time would be shorter, it was a good plan and would work, or the program was proactive and preventative. Next were those (9.4%) who expressed satisfaction that others, such as grandchildren or people living in the area, would benefit from the program.

The reasons coded in the category *prevent possible physical harm to respondent or others* (8.0%) usually involved a desire to avoid the possibility that they or others would need to worry about eating contaminated food or swimming in contaminated water. We had anticipated that some respondents would be concerned about human health and hence the first reconsideration question, B-3, addressed this concern and emphasized that the only outcome of the program would be the prevention of harm to birds, small animals, and saltwater plants. But at B-2, only six respondents gave a reason to vote *for* the program that related to possible physical harm and did not also give another reason. Furthermore, respondents who had expressed a concern about possible physical harm to humans at B-2 were significantly more likely to change their votes *for* to votes *not-for* at B-3 compared to other respondents who voted *for* ($p < 0.001$).[19] This result suggests that respondents paid attention to the information conveyed in B-3.

Among the other types of reasons given at B-2 were expressions of personal or collective responsibility to do something about the harm from oil spills because it is caused by humans (6.2%), expectations that the program would make the oil companies more responsible (4.5%), that the program might help other animals (4.4%), and that it may prevent possible permanent harm (2.5%). Responses coded as *protect environment* (2.0%) are those for which the respondent did *not* give any other type of reason (that was not coded as *other*) in their answer to B-2. Other respondents giving answers that would have been coded in this category clarified their initial response with more specific reasons in response to the non-directive probes.[20]

[18] Compared to others who also voted *for* the program, those respondents who gave this response were significantly more likely to report at D-4 that their households liked to identify different species of birds ($p = 0.009$) and to report at D-5 that their households went to the beach more than 3 times over the past summer ($p = 0.072$).

[19] Eighteen respondents changed their vote at B-3; *see* section 5.4 for a more detailed discussion.

[20] For example, in the following response, the respondent elaborated on what he or she meant by "help the environment" in response to interviewer probing: "Because it would help the environment." (PROBE) "The program will keep the shorelines clean and save the birds at the coast." (PROBE) "I want my children to be able to enjoy the shoreline as I did when I was young." See Appendix E.

5.2.3.3. *Reasons for Choosing to Vote* Against *the Program*

B-4 was asked only of those respondents who voted *against* the program:

> Did you vote against the program because it isn't worth that much money to you, or because it would be somewhat difficult for your household to pay that much, or because of some other reason?

As noted in Chapter 3, this way of asking about the respondent's motivation to vote *against* the program can alleviate the discomfort some respondents might feel at revealing motivations they find unpleasant or too personal (*e.g.*, they couldn't afford to pay for the program). Overall, 19.3 percent of those who answered this question chose the *somewhat difficult to pay* response (see Appendix C). As one would expect, the likelihood of this response was strongly related ($p < 0.001$) to the tax amount the respondents were asked about in B-1; almost five times as many respondents at the two higher tax amounts (*i.e.*, \$120 and \$220) said they could not afford it than at the two lower amounts (*i.e.*, \$5 and \$25).

Eleven percent chose the first of the offered response categories – *isn't worth that much money*; and 73.9 percent said they were against the program for some other reason (albeit, as noted below, the majority of these *other* reasons were different ways of stating the two pre-coded responses).[21] If the respondent said he or she had another reason, the interviewer was instructed to probe as to what that reason was. Each of the reasons expressed in these *other* open-ended responses were coded into the categories shown in Table 5.4. In order to give a complete picture of the responses to question B-4, also included in this table are the answers to the two pre-coded categories, *somewhat difficult to pay* and *isn't worth that much money.*

Overall, 38.7 percent of those who voted *against* voiced a concern they had about the program, such as skepticism about whether it would work, or about the payment vehicle.[22] An almost equal percentage (37.7%) mentioned that the problem was not that important or that other programs were more important to them.[23] Thirty-two percent said that the cost was too high or that the tax amount was somewhat difficult to pay.[24] Only two percent wanted more information, a percentage which suggests that most respondents who voted

[21] In 13 cases, the interviewer circled more than one B-4 answer category, hence the percentages total more than 100.

[22] Respondents who expressed concern with the program or payment vehicle at B-4 were at C-3 more likely to perceive the program as only somewhat effective ($p = 0.008$) or not too effective or not effective at all ($p < 0.001$) and at C-4 to think they would have to pay the tax for more than one year ($p = 0.003$).

[23] This includes those responses that were pre-coded into *isn't worth that much money.*

[24] This includes those responses that were pre-coded into *somewhat difficult to pay.*

Table 5.4. Reasons for Choosing to Vote *Against* the Program

B-4. Did you vote against the program because it isn't worth that much money to you, or because it would be somewhat difficult for your household to pay that much, or because of some other reason?

CODING CATEGORY	PERCENTAGE [N = 486][(a)]
Concerns about program or payment vehicle.	38.68%
Problem not that important (or *isn't worth that much money*)/Other problems more important.	37.65%
Cost too high/*somewhat difficult to pay.*	31.69%
Wants more information.	1.85%
Other.[(b)]	2.47%

(a) Percentaging base is the number of respondents who answered B-4. Categories in italics were the explicit answer categories offered to respondents. Percentages total more than 100 percent as multiple responses were allowed.

(b) Only those for whom no other category was coded.

against the program thought the information provided in the interview was sufficient. Moreover, the types of responses to B-4 and further analysis of these responses strongly suggests that respondents who voted *against* the program were attentive to the object of choice and to the financial implications of voting for it and that they weighed the object of choice against other concerns when making their decision.[25]

5.2.3.4. *Reasons for Uncertainty about Program Vote*

Respondents who said at B-1 that they were not sure how they would vote were asked B-5: "Could you tell me why you aren't sure?" As shown in Table 5.5, the verbatim responses given by the 42 respondents who were asked B-5 are similar to the responses given at B-4 for voting against the program; however, these *not sure* respondents were more likely to express a desire for more information, 23.8 percent at B-5 versus 1.9 percent of the B-4 answers of the respondents voting *against* the program. Thirty-six percent of the respondents to B-5 raised a concern about the program or, most commonly, the payment plan. Equal percentages of respondents gave at B-5 the two responses most common in B-4: 16.7 percent commented either that the problem was not that important or that other problems were more important to them, and 16.7 percent commented that the cost was too high or difficult to pay. This pattern of response suggests, as found in Carson, Hanemann *et al.* (1998), that *not sure* voters tend to vote *against* when forced to make a choice between *for* and *not-for*.

[25] *See* the discussion of the construct validity equation in section 6.6.

Table 5.5. Reasons Why *Not Sure* About Program Vote

B-5. Could you tell me why you aren't sure?

CODING CATEGORY	PERCENTAGE [N = 42][a]
Concerns about program or payment vehicle.	35.71%
Problem not that important/Other problems more important.	16.67%
Difficult to pay/cost too high.	16.67%
Wants more information.	23.81%
Other (including not sure).[b]	19.05%

[a] Percentaging base is the number of respondents who gave a response to B-5. Percentages total more than 100 percent as multiple responses were allowed.
[b] Only those for whom no other category was coded.

5.3. Vote-Assumption Questions

Respondents were asked to choose between the status quo and a program to prevent the expected harm from Central Coast oil spills over the next ten years which would cost their household a specified amount in higher taxes. As the NOAA Panel pointed out, the reliability of respondents' choices depends on the degree to which they accepted or believed certain basic assumptions underlying the choice.[26] For example, to the extent that some respondents did not believe that the prevention program would be effective, their choices would tend to under-state their values for preventing the injuries presented in the scenario. Similarly, if respondents believed that the injuries prevented by the program over the next ten years would actually be more than that described, their choices would tend to over-state their values for the injuries presented in the scenario.[27]

During this project, we devoted a great deal of effort to developing a program that as many respondents as possible would perceive as effective in preventing the specific set of injuries. Given the *ex ante* nature of the prevention program and the probabilistic nature of oil spills, some divergence between the scenario and respondent beliefs was inevitable. To monitor this, the questionnaire incorporated a set of checks on respondent acceptance of certain elements of the choice. Below, we examine the responses to those questions that asked respondents what they were assuming about various aspects of the expected harm and prevention program when they voted.

A series of questions at the beginning of Section C of the survey instrument monitored respondent acceptance of several elements of the choice, including two key items – the extent of the expected harm and the effectiveness of the

26 The NOAA Panel states (with reference to what happens when respondents do not accept information of this type) "in effect they [the respondents] will be answering a different question from that being asked" (Arrow *et al.*, 1993, p. 4605).

27 In Chapter 6 we examine the likely impacts on the estimate of willingness to pay of respondent assumptions that diverged from those features described in the survey.

Table 5.6. Respondents' Perceptions About Expected Harm[(a)]

C-1: At that time, did you think the harm from oil spills in the Central Coast over the next ten years would be about the same as that shown here, or a lot more or a lot less?

Answer categories:	Same	A lot more	A lot less	Other	Not Sure
	34.50%	34.78%	15.68%	8.67%	6.37%

[(a)] Percentaging base is the number of respondents who answered C-1.[32]

program to prevent this harm. These questions asked respondents to recall what they had in mind about certain elements of the scenario when they voted. As discussed above, introspective questions present issues of design and interpretation; these questions were designed to avoid, as much as possible, respondents misunderstanding what information we were requesting.[28]

5.3.1. *Respondent Assumptions Regarding Expected Harm*

The first debriefing question, C-1, asked respondents whether they thought that "the harm from oil spills in the Central Coast over the next ten years would be about the same as that shown here [in Card H],[29] or a lot more or a lot less." Ideally all respondents should have responded that the harm would be the same. As shown in Table 5.6, roughly equal percentages (about 35 percent) of respondents said that either the harm would be about the *same* or *a lot more*; 15.7 percent thought it would be a lot less; 6.4 percent were not sure; and the remaining 8.7 percent gave a response coded as *other*.[30] As expected, those who said *a lot more* were significantly more likely to vote *for* the program ($p < 0.001$); and those who said *a lot less*, significantly less likely to vote *for* ($p < 0.001$).[31,32]

Also included in this debriefing section was a question that serves as a general check on acceptance. Question C-2 asked, "how serious did you consider this amount of harm to be?" A little more than 11% perceived the

28 For example, respondents sometimes take this type of question as an invitation to speculate about the topic of the question instead of reporting what they had been thinking at the time they decided to vote. Some respondents may react to this type of question with annoyance because they believe the interviewer is giving them a quiz to see if they had paid attention to these features earlier in the interview.

29 Card H summarized the harm described in the scenario; *see* Appendix A.

30 An examination of the *other* responses suggests that most of these respondents were not sure; *see* Appendix E.

31 In Chapter 6, we examine the effect of respondent assumptions regarding expected harm on the estimate of willingness to pay.

32 Unless stated otherwise, throughout Chapters 5 and 6, we have used the number of respondents who answered the respective question as the percentaging base. *See* Appendix C for tabulations using all possible question outcomes as the percentaging base.

Table 5.7. Respondents' Perceptions About Effectiveness of Program[a]

C-3: Did it seem to you that the prevention program I told you about would be completely effective at preventing harm from Central Coast oil spills, mostly effective, somewhat effective, not too effective, or not effective at all?

Answer categories:	Completely effective	Mostly effective	Somewhat effective	Not too effective	Not effective at all	Not sure
	6.00%	44.69%	38.78%	5.54%	2.86%	2.12%

[a] Percentaging base is the number of respondents who answered C-3.

expected harm as *extremely serious*, 32.1% as *very serious*, 35.4% as *somewhat serious*, 16.6% as *not too serious*, and 3.9% as *not serious at all*. Those who felt that the harm was either *very serious* or *extremely serious* were significantly more likely to vote *for* the program ($p < 0.001$) and those who felt the harm was either *not too serious* or *not serious at all* were significantly less likely to vote *for* the program ($p < 0.001$).

5.3.2. *Respondent Assumptions Regarding Program Effectiveness*

Another key respondent assumption examined in Section C was how effective the respondents believed the program would be in preventing the expected harm from Central Coast oil spills. Question C-3 asked: "Did it seem to you that the prevention program I told you about would be completely effective at preventing harm from Central Coast oil spills, mostly effective, somewhat effective, not too effective, or not effective at all?" As shown in Table 5.7, 6.0 percent thought it would be *completely effective*, and 44.7 percent thought that program would be *mostly effective*. Another 38.8 percent thought it would be *somewhat effective*. Only 8.4 percent held serious doubts about its effectiveness (answering either *not too effective* or *not effective at all*), and an additional 2.1 percent were not sure about its effectiveness. Given repeatedly expressed respondent concerns about the government's competence to run such a program effectively and the level of uncertainty associated with future events, incomplete acceptance is not surprising. Further, as should be expected, those respondents who did not think that the program would be *completely effective* or *mostly effective* were less likely to vote *for* the program ($p < 0.001$).[33]

5.3.3. *Respondent Assumptions Regarding Length of Payment*

Question C-4 asked respondents whether when they voted they had thought that their households would have to pay the special tax for the program "for one year or for more than one year?" Fifty-six percent said one year, 38.9

[33] In Chapter 6, we examine the effect of respondents' assumptions about the program's effectiveness on the estimate of willingness to pay.

percent said they doubted that it would be for just one year, and 5.3 percent were unsure. This level of skepticism about the representation that the special tax payment would be in place for only one year reflects the frequently cynical views about government promises expressed by participants in focus groups and pretesting (*e.g.*, one item mentioned was the voter-approved one-year sales tax increase for San Francisco earthquake repairs which was collected for several years beyond its voter-approved term). This apparent lack of acceptance is associated with lower willingness to pay for the program, a reasonable result if respondents believed the object of choice actually entailed a higher cost than was described to them.[34]

5.3.4. *Did Respondents Feel Pressured to Vote One Way or Another?*

Question C-5 asked respondents whether they felt that the interview, overall, tried to push them to vote one way or another or let them make up their own minds. Only 5.5 percent of the total sample, or 60 respondents, said that they thought the interview had tried to push them to vote one way or another; five respondents were not sure. The 60 respondents who felt pushed to vote one way or another,[35] were asked Question C-6: "Which way did you think it pushed you?" Of those 60, 32 (2.9% of the total sample) said they felt pushed to vote *for* the program, 23 (or 2.1%) felt pushed to vote *against*, and 5 respondents were either not sure about the direction or gave a response coded as *other*.[36] These respondents were asked to explain in C-6A: "What was it that made you think that?"[37] Those who said they felt pushed to vote *for* mentioned the presentation of all the information about the expected harm, particularly the harm to the birds, the fact that only the one program was described, or that the perspective of the oil companies was not presented. Those who said they felt pushed to vote *against* felt that the harm was down-played (*e.g.*, "none are endangered," "only 12,000 [birds] involved") or that the harm was made to seem minor in contrast to the reasons to vote against the program. The small percentage of respondents that said they felt pushed suggests that most respondents perceived the survey as neutral; and the split among those respondents in regard to the direction they felt pushed suggests that the survey design achieved a reasonable degree of neutrality.

Also reassuring is the relationship between the direction these respondents felt themselves to be pushed and their votes at B-1.[38] Table 5.8 shows that

[34] *See* Chapter 6.

[35] The five respondents who said *not sure* at C-5 were not asked this follow-up question but rather skipped to the next question.

[36] *See* Appendix E.

[37] *See* Appendix E.

[38] Two respondents who reported they felt pushed to vote against later changed their vote when given an opportunity to reconsider at D-15. The reconsideration questions, B-3 and D-15, are discussed below.

Table 5.8. Voting Patterns by Direction Felt Pushed

Direction Felt Pushed	Voted *For*		Voted *not-for*
Pushed *For* [N = 32]	31.25%		68.75%
Not Pushed [N = 1019]	51.62%		48.38%
		$\chi^2_{(1)} = 5.15$; p = 0.023	
Pushed *Against* [N = 23]	56.52%		43.48%
Not Pushed [N = 1019]	51.62%		48.38%
		$\chi^2_{(1)} = 0.22$; p = 0.642	

those who felt pushed to vote *against* actually voted *for* the program with virtually the same frequency (p = 0.642) as the 94 percent of the sample who said they felt the interview let them make up their own mind. Thus it appears that respondents who felt pressured to vote against were not influenced to vote for or against. In contrast, the voting behaviors of those who said they felt pushed to vote *for* the program and those who said they did not feel pushed exhibit a significant difference (p = 0.023): those who felt pushed to vote *for* voted *against* the program more often than those who felt the interview let them make up their own mind. This is the opposite result from that one might have thought *a priori*. Using a more appropriate one-sided test, which takes into account the hypothesized direction of the difference, to compare the voting behavior for those who felt pushed *for* to those who did not feel pushed results in a p-value of 0.993.[39] Similarly comparing those who felt pushed *against* to those who did not feel pushed results in a p-value of 0.401.

5.4. Reconsideration Questions

Two questions included in the survey instrument gave respondents an opportunity to reconsider and change their votes. The first reconsideration opportunity, B-3, was only asked if the respondent had voted *for* the program at B-1. B-3 directs respondents to "suppose human health was definitely not affected" and that "the program would only prevent harm to birds, small animals, and saltwater plants." B-3 then asks: "Would you vote for or against the program if it cost your household [B-1 tax amount]?" Although this question was directed toward cleansing the health concerns from their vote, it also offered respondents who voted *for* a chance to change their vote for *any* other reason. To be conservative, we counted those who said *not sure* to B-3 as having changed their vote to *against*.[40] These combined categories (*i.e.*, *against* or *not*

[39] This Fisher's exact test compares the null hypothesis of no difference to the alternative hypothesis that respondents who felt pushed to vote *for* were more likely to vote *for*. *See* Lehmann, 1986 for a discussion of Fisher's exact test.

[40] This treatment is consistent with the treatment in the next chapter of those who responded *not sure* to the initial voting question.

sure) are referred to below as *not-for* votes. The other reconsideration question, D-15, was asked much later in the interview and was asked of all respondents. Here again, those who expressed uncertainty were treated as *not-for* votes.

Four percent of those who originally voted *for* the program, a total of 22 respondents, changed their votes from *for* to *not-for*, with most ($N = 18$) switching their vote at the first opportunity offered, B-3. The respondents who changed their votes from *for* to *not-for* have certain distinguishing characteristics when compared to those who did not change. Respondents who did not pay taxes (and for whom the tax payment may not have been incentive-compatible) were much more likely to change from *for* to *not for* ($p = 0.034$); and those in the lowest three income categories were almost twice as likely to change to *not-for* ($p = 0.015$). Another category of respondents who were more likely to switch from *for* to *not for* are those who the interviewer identified at E-5 or E-6 as having some difficulties understanding either the harm, the program, or the voting question ($p = 0.006$).

In past studies, we have usually asked vote reconsideration questions only of respondents who voted *for* a program to avoid the possibility that respondents not voting for a program might feel pushed to vote for the program. In this study, for research purposes, the second reconsideration question, D-15, was asked of everyone; but we continued our usual practice of using for estimation purposes only the votes of those who changed their votes from *for* to *not-for*.

At D-15, 42 respondents, 7.9 percent of those who originally voted *not-for* or 3.9 percent of the total sample, changed their votes to *for*. Respondents who changed from *not-for* to *for* were significantly more likely to have said *not sure* to the first choice question, B-1 ($p < 0.001$). Twelve (28.6 percent) of those who changed from *not-for* to *for* were unsure at B-1. The other notable characteristic of this group was that those who changed from *not-for* to *for* were significantly more likely ($p < 0.001$) to have lower levels of schooling (high school diploma or less) than the rest of the sample. While it may be that some respondents who changed from *not for* to *for*, particularly those who were initially *not sure* rather than against at B-1, were truly willing to pay the specified tax amount, the significant relationship of this group with low education suggests some of them may have felt prompted to change their vote.

5.5. Interviewer-Evaluation Questions

The answers to the series of questions in Section E that the interviewers answered after leaving the respondents' homes are another source of information about whether respondents understood the voting choice.[41] Although

[41] The first three questions in Section E asked the interviewer to record the respondent's sex, race, and zip code.

Table 5.9. Interviewer Evaluation of Respondent Reaction to Choice Elements[a]

E-4 Questions	Extremely	Very	Somewhat	Slightly	Not at all	Not Sure
How distracted?	0.28%	1.29%	6.93%	18.93%	72.48%	0.09%
How attentive?	24.65%	61.59%	13.30%	0.37%	0.09%	0.00%
How interested?	22.85%	51.16%	22.66%	2.13%	0.93%	0.28%

[a] Percentaging base is the number of interviewers who answered each question.

interviewer evaluations are necessarily subjective, the interviewers were experienced in-person interviewers, able to closely observe the respondent during the interview. Our training emphasized that interviewers should be frank in making these appraisals.[42]

Question E-4 asked the interviewers to assess the respondent's reactions to the portion of the interview that presented the scenario, including the extent of the expected harm and the prevention program. Table 5.9 shows the interviewer ratings from E-4 for how distracted, attentive, and interested the respondent appeared during this part of the presentation. As shown in the table, very low percentages of respondents were rated as *extremely* or *very distracted* (1.6%) during the presentation; and a large majority, 72.5 percent, were rated as *not at all distracted*. Less than one percent were rated by the interviewers as only *slightly* or *not at all attentive*, and 86.2 percent were rated as either *very* or *extremely attentive*. A very low percentage was rated as *not at all* or *slightly interested* (3.1%); most were rated as either *extremely* or *very interested* (74%). Those who were rated as either *very or extremely distracted* or *slightly or not at all attentive* or *interested*, a total of 47 respondents, were significantly less likely to vote *for* the program ($p < 0.001$).

Question E-5 asked the interviewer if the respondent had said anything that suggested difficulty understanding either the harm caused by oil spills or the prevention program. A total of 48 respondents (or 4.42% of the total sample) were identified as having had a difficulty of some sort. In the open-ended question E-5A, the interviewers were asked to describe the difficulties.[43] As described by the interviewers, these respondents had either difficulty seeing the visual aids or misunderstood an aspect of the expected harm or program (which was sometimes subsequently clarified). Some interviewers reiterated respondents' questions about aspects of the harm or the program that were asked and recorded earlier in the interview. With respect to the vote questions, respondents whom the interviewers identified as having a problem understanding these aspects of the interview were not significantly different from other respondents ($p = 0.451$); these respondents were however more likely to change their vote from *for* to *not-for* the program ($p = 0.002$).

The next two questions asked about the respondent's reaction to the choice

[42] Westat, 1995, pp. 4-3 and 4-69.

[43] *See* Appendix E.

Table 5.10. Interviewer Evaluation of the Seriousness of Respondent Consideration of the Voting Decisions[(a)]

Question E-8	Extremely	Very	Somewhat	Slightly	Not at all	Not Sure
How serious was the consideration the respondent gave to the decision about how to vote?	24.95%	55.57%	18.00%	0.93%	0.28%	0.28%

[(a)] Percentaging base is the number of interviewers who answered E-8.

question, B-1. Question E-6 asked if the respondent had any difficulty understanding B-1; and, if so, the interviewer was asked at E-6A to describe the difficulties.[44] A total of 19 respondents (1.8% of the total sample) fell into this category. Their difficulties, as described by the interviewers, included respondent skepticism about the length of time of the tax payment or about another aspect of the payment plan. Some of these difficulties appear to have been overcome after the interviewer reread the pertinent material or question.

Another factor that might affect a respondent's understanding of the choice is impatience to get through the interview. Questions E-7 and E-7A asked the interviewer to rate the degree of impatience the respondent displayed when asked the voting questions. The vast majority of the respondents (84.9%) were not thought to be impatient, and another 6.2 percent were rated as *not very impatient* or only *a little impatient.* Four percent were said to be *somewhat impatient*, and 1.7 percent were said to be *very impatient.* Those who were considered by the interviewer to be *very* or *somewhat impatient* were significantly less likely to vote for the program ($p < 0.001$).

Interviewer ratings in response to E-8 – "How serious was the consideration the respondent gave to the decision about how to vote?" – can be used to gauge whether the choice mechanism was plausible and taken seriously by respondents. As shown in Table 5.10, 80.5 percent of the total sample were thought to have given the matter *extremely* or *very serious consideration.* Only about 1.2 percent (or 13 cases) were rated as giving it *not at all* or only *slightly serious consideration*; and these 13 respondents were significantly less likely to vote for the program ($p = 0.015$).

5.5.1. *Were Respondents' Choices Influenced by Others?*

In order to avoid distractions, interviewers were instructed to try to conduct interviews without other persons present. Frequently, any other people were present were young children in the respondent's care. In order to differentiate

[44] *See* Appendix E.

these cases from those where teenagers or adults were present, the interviewers were asked in E-9 to report whether anyone age 13 or older was present when the respondent voted. In 21 percent of the interviews, someone age 13 or older was present.

In question E-9A, the interviewer was asked whether the other person(s) seemed to affect how the respondent voted. In almost 90 percent of the cases where someone age 13 or older was present while the respondent voted, the interviewers judged that there was no effect. In six cases (less than 1% of the sample), the interviewer believed that the other person present did have an effect; and in 18 cases the interviewers indicated that they did not know whether the other person affected the way the respondent voted. For these 24 cases we examined the interviewer responses to E-9A and E-10: whenever the interviewer mentioned possible influence, it was almost always by another household member.[45] Since the goal of the survey was to measure *household* preferences, any influence of other household members in this small number of cases would not be inconsistent with that goal.

5.6. Summary

The pattern of responses to the open-ended questions considered in this chapter was consistent with respondents paying attention to the survey and taking the choice opportunity seriously. Answers to questions about the reasons for their voting choices (B-2, B-4, and B-5) generally referred to relevant features of the prevention program such as what the program would accomplish and its cost. Overall, the types of queries raised during the presentation of the expected harm and the prevention program were usually related to the material in a meaningful way and provide assurance that the respondents were paying attention to this part of the interview. Further, coupled with the responses to the Section C vote-assumption questions and the Section E interviewer-evaluation questions, these patterns suggest that respondents' decisions reflected their perceptions of the object of choice and their preferences for it.

An important feature of our design was to offer respondents opportunities to reconsider their choices. Question B-3 allowed those who voted *for* the program to reconsider their votes shortly after they voted; and another reconsideration question, D-15, was offered to all respondents near the end of the interview. Those who gave a response to B-2 related to *possible effects on humans* were more likely to reconsider and change their *for* vote to an *against* vote compared to the rest of the sample.

Analysis of the interviewer-evaluation questions in Section E of the survey

[45] *See* Appendix E.

found that in only a few cases the interviewers identified possible problems with respondents' attentiveness, interest, and impatience. Further, those whom the interviewers identified as less focused were generally less likely to vote for the program compared to the rest of the sample.

CHAPTER 6
Analysis of Choice Questions

6.1. Introduction

In this chapter the choices made by respondents in the main survey are used to construct a lower-bound estimate of the *ex ante* total value for preventing the expected harm from oil spills along the California Central Coast over the next decade. The relationships between the choice measure and other respondent characteristics measured by the survey are also examined. Section 6.2 presents two versions of the choice measure. Section 6.3 discusses the non-parametric (Turnbull, 1976) statistical framework used in much of our analysis of the estimate of *ex ante* total value. Section 6.4 provides the Turnbull lower bound estimate on the sample mean[1] and examines the sensitivity of this estimate to various assumptions regarding the treatment of the data. Using the categories suggested by the NOAA Panel[2] as a framework, section 6.5 examines the bivariate relationships between choice measures and respondent characteristics. Section 6.6 examines construct validity using a multivariate counterpart to the evaluations of individual variables reported in the prior section. Section 6.7 provides a sensitivity analysis that looks at possible shifts in value related to respondent assumptions at variance with key scenario features. Finally, section 6.8 presents the most conservative treatment of the respondents who said that they did not pay California income taxes and its impact on the total value estimate.

6.2. Definition of Choice Measures

B-1, the principal choice question, asked respondents if they would vote for or against a Central Coast prevention program if it cost their household a certain tax amount. Before survey administration, the survey research company had randomly assigned each respondent to one of five different tax amounts (referred to as B1AMT below): \$5, \$25, \$65, \$120, or \$220. Table 6.1a summarizes the responses to the B-1 choice question by B1AMT. In the following analysis, the *against* and *not-sure* categories of question B-1 (displayed in the last two

[1] *See* Appendix F for a more detailed discussion of the Turnbull estimator and Appendix I for a comparative analysis of COS and the *Exxon Valdez* survey results.

[2] Arrow *et al.*, 1993, p. 4609.

Table 6.1a. B-1 Responses by B1AMT

B1AMT	For	Against	Not Sure[3]
$5	69.86%	27.40%	2.74%
$25	58.33%	39.35%	2.31%
$65	51.04%	45.23%	3.73%
$120	45.30%	49.17%	5.52%
$220	29.82%	64.47%	5.70%

Table 6.1b. **B1** Choice Measure by B1AMT

B1AMT	For	Not For
$5	69.86%	30.14%
$25	58.33%	41.67%
$65	51.04%	48.96%
$120	45.30%	54.70%
$220	29.82%	70.18%
	$\chi^2_{(4)} = 79.08$; $p < 0.001$	

columns of Table 6.1a) are combined into a single *not-for* category; the choice measure incorporating this coding will be referred to as **B1**.[4]

The key prediction of economic theory is that the percentage of the sample voting *for* should decrease as the tax amount increases. Table 6.1b displays the percentages of *for* and *not-for* responses to **B1** by B1AMT. Based on these data, a chi-squared test ($\chi^2(4)$[5] $= 79.08$; $p < 0.001$) clearly rejects the null hypothesis that the percentage *voting-for* does not systematically vary with B1AMT. A one-sided Fisher's exact test that takes into account the theoretically predicted direction of the variation of **B1** with B1AMT provides an even stronger rejection of the null hypothesis.

A choice measure defined only by the B-1 responses (*e.g.*, the **B1** choice measure defined above) results in *single-bounded* interval data.[6] That is, if a respondent votes *for*, the respondent's willingness to pay (WTP) for the program is bounded from below by B1AMT (*i.e.*, the respondent is willing to pay at least B1AMT). How much more the respondent might be willing to pay is not

3 In a $65 version questionnaire, the interviewer did not circle an answer category at B-1. Given the nature of the comments recorded verbatim by the interviewer at B-1 ("*I'm not going to answer that ... can't say for or against*"), the B-1 response for this case was coded as *not sure.*

4 All choice measure variables are denoted in bold capital letters.

5 The null hypothesis tested in a chi-squared (χ^2) test is that the rows and columns in a two-way table are independent.

6 The seminal paper on the use of binary discrete choice data in contingent valuation is Bishop and Heberlein (1979). Hanemann (1984) developed the utility-theoretic approach to such models. Cameron and James (1986) look at choice data using an approach based on the willingness-to-pay function. McConnell (1990) compares the two approaches.

Table 6.2. **B1CH** Choice Measure by B1AMT

B1AMT	For	Not For
\$5	68.95%	31.05%
\$25	56.94%	43.06%
\$65	48.55%	51.45%
\$120	40.33%	59.67%
\$220	28.95%	71.05%
	$\chi^2_{(4)} = 82.48$; $p < 0.001$	

revealed; we know only that the respondent's WTP is not less than B1AMT ($WTP_R \geq$ B1AMT). If the respondent gives a *not-for* answer, the respondent's willingness to pay is bounded from above by B1AMT (*i.e.*, the respondent may be willing to pay some tax amount below B1AMT or may not be willing to pay anything at all). Thus, the respondent's WTP is less than B1AMT ($0 \leq WTP_R <$ B1AMT).[7]

6.3. Statistical Framework for Analysis

In developing an estimate of the *ex ante* total value for preventing the expected harm from oil spills off California's Central Coast over the next ten years, we have consistently chosen conservative design features and statistical assumptions.[8] Respondents who voted *for* the program at B-1 were allowed to reconsider that vote in question B-3 and again in question D-15.[9] Revising the **B1** choice measure (defined above) to take into account those respondents who changed their votes *for* to votes *not-for* results in a second choice measure – **B1CH**. **B1CH** treats as votes *for* only those respondents who voted *for* the program at B-1 and who did not change their votes *for* at either of the two opportunities (*i.e.*, at B-3 and D-15); by construction, **B1CH** is a more conservative choice measure than **B1**.

Table 6.2 displays the **B1CH** choice measure by B1AMT; the null hypothesis that the **B1CH** choice measure does not systematically vary with B1AMT is rejected $\chi^2(4) = 82.48$, $p < 0.001$). Unless otherwise indicated, we use the **B1CH** choice measure for the remaining analysis presented in this chapter.

To estimate *ex ante* total value, we have chosen as our summary statistic the Turnbull (1976) non-parametric, maximum likelihood (ML) estimator for

7 We assume that no respondent would demand compensation for implementing a program to prevent oil spills along the Central Coast, *i.e.*, no respondent has a negative WTP.

8 This is in keeping with the NOAA Panel's recommendation: "Generally, when aspects of the survey design and the analysis of the responses are ambiguous, the option that tends to underestimate willingness to pay is preferred" (Arrow *et al.*, 1993, p. 4612).

9 Respondents who voted *against* the program at B-1 were also given an opportunity to reconsider their vote at D-15.

interval-censored data.[10] The Turnbull estimator uses respondents' choices to construct an interval estimate for the latent willingness to pay implied by each respondent's choice. As described above, an individual's choice will distinguish either a lower or an upper bound for his or her WTP. By combining respondents' choices, we obtain estimates for the relative frequency of responses at different WTP intervals, (0, $B1AMT_i$) and ($B1AMT_i$, ∞), where $B1AMT_i$ is one of the five B-1 tax amounts administered to independent sub-samples. The first pair (0, $B1AMT_i$) defines the interval with $B1AMT_i$ as an upper bound; and the second pair ($B1AMT_i$, ∞), the interval with $B1AMT_i$ as a lower bound. The six intervals or *steps* defined by B1AMT are (1) \$0–\$5, (2) \$5–\$25, (3) \$25–\$65, (4) \$65–\$120, (5) \$120–\$220, and (6) \$220–∞.

A range of summary statistics related to the sample mean can be defined based on the Turnbull estimates of the fraction of the sample in each of the six intervals. These estimates differ in the assumed distribution of respondents among the six intervals. The lowest of these we will refer to as the lower bound on the sample mean. The fraction of the sample estimated to be in each interval is treated as having a willingness-to-pay value equal to the lower end-point of the interval, and then the ordinary sample mean is calculated.[11] The highest of these summary statistics is the upper bound on the sample mean. The fraction of the respondents estimated to be in an interval is treated as having a WTP at the high end-point of the interval, and then the ordinary sample mean is calculated.[12]

Irrespective of the particular tax amounts used to define the intervals, the unobserved sample mean is always bounded below by the lower bound on the sample mean and above by the upper bound on the sample mean, if there are equivalent sub-samples at each of the tax amounts.[13] However, the particular tax amounts respondents are asked about influence how much *less* the Turnbull estimate of the lower bound on the sample mean is than the sample mean and how much *greater* the Turnbull estimate of the upper bound on the sample mean is than the sample mean. Any estimate of the sample mean which is

[10] The initial uses of this framework in the CV literature are found in Carson and Steinberg (1990) and Kristrom (1990). Haab and McConnell (1997; 2002) provide further development. See Carson, Willis, and Imber (1994) for a large-scale application. Appendix F provides a detailed discussion of the Turnbull estimator.

[11] For example, if 20% of the sample is estimated to be in the interval \$25 to \$65, the lower-bound mean is calculated by assuming that this 20% of the sample is willing to pay exactly \$25.

[12] For example, if 20% of the sample is estimated to be in the interval \$25 to \$65, the upper-bound mean is calculated by assuming that this 20% of the sample is willing to pay \$65. As the high end-point in the last interval, \$220–∞, is infinity, the upper-bound mean is infinite unless reasonable additional assumptions are imposed. As we are asking about WTP, it would be possible to substitute for infinity an upper bound based on either income or wealth. *See* Section 2 in Appendix F.

[13] In large but finite random samples such as the one used for this study, the number of respondents receiving each tax amount is approximately equivalent. The standard error of the estimate reflects the sampling variation.

Table 6.3. Turnbull Estimate of WTP Distribution and Lower Bound on the Sample Mean: **B1CH** Choice Measure [N = 1,085]

Lower Bound of Interval	Upper Bound of Interval	Probability of Voting *For* at Upper Bound	Change in Density[14]
$0	$5	0.6895	0.3105
$5	$25	0.5694	0.1201
$25	$65	0.4855	0.0840
$65	$120	0.4033	0.0822
$120	$220	0.2895	0.1138
$220	∞	0.0000	0.2895

Log-Likelihood = −709.48
Estimate of lower bound on sample mean = $85.39
Standard error of the estimate = $3.90

lower than the Turnbull estimate of the lower bound on the sample mean or higher than the Turnbull estimate of the upper bound on the sample mean is inconsistent with the observed choices made by respondents. Without additional statistical assumptions about the latent willingness-to-pay distribution, any other observed choice measure is uninformative about where, within the two Turnbull bounds, the sample mean lies. The most conservative assumption consistent with the observed choices is that the sample mean is equal to the Turnbull estimate of the lower bound on the sample mean.

6.4. Turnbull Estimate of the Lower Bound on the Sample Mean

Table 6.3 reports the Turnbull estimate of the lower bound on the sample mean for the WTP distribution using the **B1CH** choice measure. Note that the third column in the table (labeled "Probability of Voting *For* at Upper-Bound") is simply the estimated fraction of those in Table 6.2 who would vote *for* the program at each B1AMT. Table 6.3 describes the intervals defined by B1AMT and respondent choices. For example, we know a respondent's willingness to pay for the Central Coast prevention program is greater than or equal to $5 if the respondent voted *for* the program at $5. If, on the other hand, a respondent voted *against* the program at $5, we know that the respondent's willingness to pay is less than $5 and possibly $0. Likewise, for a respondent who was asked

[14] The values shown in the change in density column are the percentage of respondents who fall into each interval. For example, 12.0 percent of respondents fall into the $5–$25 interval; and, hence, the Turnbull assumes 12.0 percent are willing to pay $5. The z-statistics for the five change in density parameters estimated by the model are 9.93, 2.61, 1.80, 1.69, and 2.41, respectively. The significance of each individual parameter value is of little importance; the set of parameters taken together, however, is reflected in the standard error of the estimate. The standard error of $3.90 suggests reasonable precision in the estimate.

about $25, a vote *against* the program implies that the respondent's willingness to pay for the prevention program lies somewhere in an interval from $0 to $25; while a vote *for* implies a minimum willingness to pay of at least $25. In this way, we can classify each respondent's willingness to pay into an interval depending on the B1AMT the respondent received. The Turnbull estimate of the lower bound is then calculated by multiplying the lower bound of the interval column by the change in density column and then summing the products.[15]

The Turnbull estimate of the lower bound on the sample mean of $85.39[16] is obtained by assuming that all of the fraction of the sample estimated to be in a particular interval falls at the lower end of that interval. For example, respondents who voted *against* at $5 fall into the $0–$5 interval and are assumed to have a willingness to pay of $0. Respondents who voted *for* at $220 fall into the $220–∞ interval and are assumed to have a willingness to pay of $220. The median (50th percentile) respondent falls in the $25–$65 interval.

6.4.1. *Sensitivity of the Turnbull Estimate of the Lower Bound*

In this section we examine the sensitivity of the Turnbull estimate of the lower bound on the sample mean to the exclusion from the estimate of nine categories of respondents (summarized in Table 6.4 below).

The first category of respondents are those who were unsure about how they would vote on the program. In the prior section, these 43 respondents were treated as *not-for* the program. A less conservative treatment is to not include them in the estimation. Not including these observations raises the Turnbull estimate of the lower bound on the sample mean by $4.16, from $85.39 to $89.55.

Several categories consist of respondents who *protest* the payment mechanism, *i.e.*, they object that the oil companies, not the taxpayers, should be paying for the prevention program. The two most common locations at which this sentiment was spontaneously expressed by respondents are, first, before the choice question (*e.g.*, during the presentation of the payment vehicle) and, second, in response to the choice question. Some respondents also protested in their responses to the open-ended vote-motivation questions (B-2, B-4, and B-5) or spontaneously at later points in the interview. Excluding the 7.1 percent who protested before the choice question results in a $2.32 increase in the lower bound on the sample mean. Excluding, in addition, those who protested at the choice question results in a slightly higher increase of $3.87. Using the most inclusive definition of protester and excluding the 15.6 percent who

[15] *See* Appendix F.

[16] The Turnbull estimate of the lower bound on the sample mean using the sample weights (see Appendix B.10) is $85.50, only $0.11 higher than the unweighted sample estimate. The standard error of the weighted estimate is $3.84.

Table 6.4. Summary of Sensitivity Tests

Part of Sample Not Included	Percentage of Sample Not Included In Estimation	Change in $85 **B1CH** Lower-Bound Estimate
Not sure responses to choice question [B-1]	3.96%	+$4.16
Protested that oil companies should pay before B-1	7.10%	+$2.32
Protested that oil companies should pay before or during B-1	10.32%	+$3.87
Protested that oil companies should pay at any point during interview	15.58%	+$8.74
Protested that oil companies would pass cost on to consumers at any point during interview	5.35%	+$3.24
Felt pushed to vote one way or another [C-5]	5.07%	+$0.43
Not at all/slightly serious consideration of B-1 [E-8]	1.20%	+$0.74
Negative evaluations by interviewer on one of six indices (includes R's in previous category)	9.95%	−$0.33
WTP more than 5% of income	0.37%	−$1.27

protested that the oil companies should pay for the program at any point in the interview results in an increase of $8.74.

Another category consists of protesters that objected that oil companies would pass their share of the program costs (*i.e.*, operating costs over the next 10 years) to the consumer in the form of higher prices. Not including the 5.4 percent of the sample who expressed this view at any point in the interview results in an increase of $3.24 in the estimate. This group shares some overlap with those who protested that the oil companies should pay.

Another category of respondents are those who felt pushed to vote one way or another.[17] Not including the 5.1 percent who felt pushed to vote either *for* or *against* the program results in an increase of $0.43.

The interview evaluation questions in Section E can be used to identify respondents who may have had problems understanding an element of the survey. The most obvious group is that of respondents who the interviewers said gave the choice question *slightly serious* or *not at all* serious consideration (E-8). Not including this 1.2 percent of the sample raises the estimate of the lower bound on the sample mean by $0.74. A more expansive definition also encompasses respondents the interviewers identified either as having some difficulty understanding the harm caused by Central Coast oil spills or the prevention program (E-5), as having some difficulty understanding the choice question (E-6), or as being *very* or *extremely* distracted (E-4a), *slightly* or *not at all* attentive (E-4b), or *slightly* or *not at all* interested (E-4c) during the

[17] *See* section 5.3.4.

presentation of the descriptive material (*i.e.*, A-3 through A-13; see Appendix A). Not including this 10 percent in the sample results in a \$0.33 decrease in the lower bound on the sample mean. This analysis suggests that those who did not take the choice seriously, who had any difficulties understanding, or who were distracted, inattentive, or uninterested were willing to pay slightly less than other respondents in the sample.

Respondents voting to pay a large percentage of their income is an indication that they may not have taken their budget constraints seriously. Taking the ratio of the lower bound of the interval where the respondent's willingness-to-pay amount lies to the respondent's income, we find that no respondents are willing to pay more than 10 percent of their household income; and only four respondents, all of whom reported that they did not pay state income taxes, are willing to pay more than 5 percent. Excluding these four respondents from the estimation results in a decrease of \$1.27 in the estimate.[18] We consider a possible adjustment for respondents not paying state income taxes in section 6.8 below.

6.4.2. *Tests for Nay-sayers and Yea-sayers*

The Turnbull non-parametric estimator fit to the **B1CH** choice measure allows for the possibility that some fraction of the respondents might vote *not-for* even if the cost to the respondent was \$0; and it also allows for the possibility that some fraction of the respondents might vote *for* the program irrespective of the tax amount. Such respondents are often referred to as nay-sayers and yea-sayers in the social science and contingent valuation literature (Mitchell and Carson, 1989; Hanemann and Kanninen, 1999) and as *natural mortality* and *immunes* in the biometrics literature where statistical methods of dealing with these phenomena were first developed (Finney, 1949).

Figure 6.1 depicts an illustrative distribution of which 60% of the respondents react to the amount they are asked to pay in a linear fashion; 20% of the respondents are nay-sayers; and 20% of the respondents are yea-sayers. The nay-sayers are represented as a *vertical spike* at zero, while the yea-sayers are represented as a *horizontally-oriented slice* at the bottom of the graph. The regression line now effectively starts at 80% *for* at \$0.01 and ends with 20% being willing to pay an arbitrarily large amount.

The presence of nay-sayers or yea-sayers can influence the interpretation of

[18] Eighteen respondents violate a strict two percent of income criterion and 30 violate a very strict one percent criterion. These respondents are significantly more likely to have reported they did not pay income taxes ($p < 0.001$).

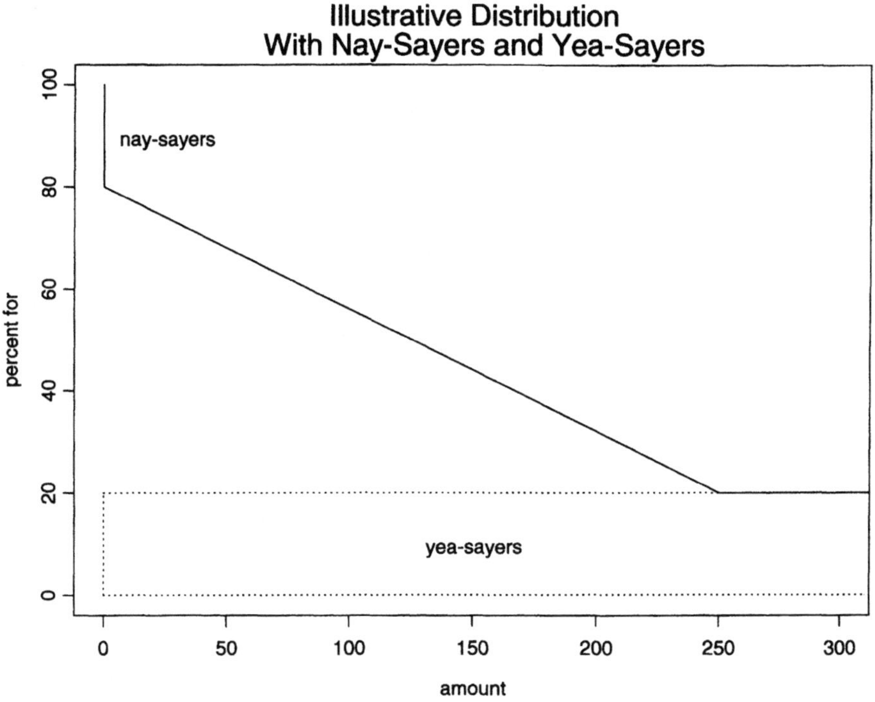

Figure 6.1.

the Turnbull lower bound on the sample mean.[19] Due to its non-parametric nature, the Turnbull approach is incapable of providing estimates of either the fraction of nay-sayers or the fraction of yea-sayers. To obtain such estimates, it is necessary to make a parametric assumption concerning the distribution of the underlying WTP distribution. The standard parametric distributions for survival analysis (Nelson, 1982) do not allow for the possibility of either nay-sayers or yea-sayers. One may include parameters allowing for the possibility of nay-sayers, yea-sayers, or both and then test whether the fit of the parametric model is improved.

To perform such a test for nay-sayers, we consider three commonly used parametric models: the log-normal, the Weibull, and the Box-Cox. The log-normal and the Weibull are two-parameter distributions. Adding a parameter allowing for the possibility of a spike at zero (*i.e.*, nay-sayers) in the log-normal model results in an estimate that 28.7% of the respondents have a zero WTP; and in the Weibull model, an estimate that 24.2% of the respondents have a

[19] True zeros are correctly taken into account in the calculation of the Turnbull lower bound on the sample mean, while nay-sayers who do have a positive WTP for the good under the scenario depicted will bias the Turnbull lower-bound on the sample mean downward. The presence of yea-sayers will bias the Turnbull lower bound on the sample mean upward.

zero WTP. The individual z-statistics on the spike parameters (4.75 and 2.61, respectively) are both significant at the $p < 0.01$ level, but the likelihood ratio test for the log-normal model with and without the naysayer-parameter $\chi^2(1) = 2.99$, $p = 0.08$) and that for the Weibull model with and without the naysayer-parameter $\chi^2(1) = 1.73$, $p = 0.19$) suggest that these two distributions do not fit dramatically better as a result of adding the naysayer-parameter.[20]

The Box-Cox is a three-parameter distribution that nests the log-normal and normal as special cases. The three-parameter Box-Cox model allows more flexible curvature with respect to the shape of the underlying WTP distribution than either of the two-parameter models above, the log-normal and the Weibull. The estimate of the Box-Cox λ parameter of 0.3665 suggests a distributional shape between that of the normal ($\lambda = 1$) and that of the log-normal ($\lambda = 0$). Likelihood ratio tests reject the normal in favor of the Box-Cox at $p = 0.01$ and the log-normal in favor of the Box-Cox at $p = 0.07$. An examination of the Box-Cox fit in this case shows the model predicting a steep drop in the percentage willing to pay as one moves away from zero. As a result, the inclusion of a zero spike parameter does not improve the statistical fit of the Box-Cox model to the data. Allowing for the possibility of a spike at zero in the Box-Cox model results in an estimate of the fraction of respondents at zero of less than 0.1%.

Thus, while the nay-sayer tests based on the two-parameter distributions support the possibility of a downward bias in the Turnbull lower bound on the sample mean, the Box-Cox model provides an equally good fit to the data while allowing more flexibility in the percentage of the respondents who are estimated to have very small, but still positive, WTP values. Comparison of the log-normal with a spike, the Weibull with a spike, and the Box-Cox (all three-parameter distributions) to the Turnbull (which is a perfect fit to the data) suggests that all three of the parametric distributions fit the data quite well.[21]

Turning now to the possibility of *yea-sayers*, all three parametric models suggest the same answer: the fraction of respondents that are yea-sayers is less than 0.1%. The yea-sayer parameter never approaches significance in any of the three models. Further, the log-likelihood of each of the three models with the yea-sayer parameter is almost identical to that of the corresponding model without the yea-sayer parameter: a result which suggests that allowing for the possibility that some respondents are willing to pay any amount provides almost no improvement in the distributional fit.

Allowing for the possibility of both nay-sayers and yea-sayers at the same

[20] The log-likelihoods for the log-normal model and the log-normal model with a spike at zero are −711.31 and −709.81, respectively. For the Weibull and Weibull spike models the log-likelihoods are −710.54 and −709.67, respectively.

[21] The log-likelihoods for the log-normal with a spike at zero, the Weibull spike model, the Box-Cox model and the Turnbull are −709.81, −709.67, −709.63, and −709.48, respectively.

time does not change this conclusion. If the possibility of both nay-sayers and yea-sayers is allowed in the log-normal and Weibull distributions, the nay-sayer parameter again suggests 20–30% of the observations should be placed in a spike at zero. The yea-sayer parameter estimates are again very close to zero. In the Box-Cox model with nay-sayer and yea-sayer parameters, both of those parameter estimates are essentially zero and insignificant. In all three cases, the resulting distributions look equivalent to the models that only allowed for nay-sayers. Thus, none of the tests performed with both parameters provide any support for the presence of yea-sayers who might bias the Turnbull lower bound on the sample mean upward; nor do they provide any ground for choosing as the correct model either a spike of nay-sayers at zero or a substantial number of small but non-zero WTP values.

6.5. Bivariate Relationships Including Cross-Tabulations Recommended by NOAA Panel

The NOAA Panel recommends categorizing the responses to the primary valuation question (*e.g.*, in this survey, the **B1** and **B1CH** choice measures) to facilitate interpretation of the responses to this question. The recommended categories include income, prior knowledge of the site, variables related to prior interest in the site, distance to the site, attitudes toward the environment, attitudes toward big business, understanding of the task, beliefs about the scenario, and ability or willingness to perform the task (Arrow *et al.*, 1993, p. 4609). First, we report the cross-tabulation results for variables reflecting each of the recommended categories with the choice measures **B1** and **B1CH**. Second, for illustrative purposes, for a subset of these recommended categories, we consider the way the Turnbull estimate of lower bound on the sample mean varies according to the measurement of these categories. Third, in Section 6.6, we present a multivariate analysis addressing a subset of the categories recommended by the Panel and others that economic theory suggests should influence respondents' choices.

Table 6.5 describes the specific source of the information used in each of the cross-tabulations. In most cases, the variables are constructed from single questions in the main study survey instrument. In a few cases, we constructed the measure using two or more questions. The table also includes a short descriptive summary of the information and an indication of whether the variable directly (*D*) or indirectly (*I*) measures the category identified by the Panel. As shown by the table, the survey instrument contains multiple variables for some of the Panel's recommended categories.

Table 6.6 summarizes the cross-tabulation results for the **B1** and **B1CH** choice measures.[22] For each cross-tabulation, we test the null hypothesis that

[22] Appendix G presents cross-tabulation tables for each of the variables in Table 6.6 by the **B1** and **B1CH** choice measures.

Table 6.5. Description of Sources of Information for Cross-Tabulations

Recommended Category	Source	Relationship to Category	Description
1. Income	D-12	D	Total household income before taxes in 1994
2. Prior Knowledge of Site	A-7 D-1	D	Familiar with 5 types of birds affected in spills; Driven along the Central Coast on Highway 1
3. Prior Interest in the Site	A-4 D-5	I	Visited any of three types (beach, marsh, rocky shore) of California shoreline in last twelve months; Visited beach at least three times last summer
	D-3 D-4	I	Saltwater boating or fishing in the last 5 years; Bird watcher
4. Attitudes Toward the Environment	A-1b A-1e A-2c	D	Reducing air pollution in California cities; Protecting coastal areas from oil spills; Protecting wildlife
	D-7	I	Respondent's self-evaluation on environmentalist scale
5. Attitudes Toward Big Business	Oil companies should pay	I	Oil companies should pay for all of program costs (*i.e.*, Box 4 or Box 5 checked by interviewer)
6. Distance to the Site	Central Coast PSU's	D	Location of respondent's residence in Central Coast PSU's (*i.e.*, San Francisco Bay down to greater Los Angeles area)
7. Understanding of the Task	E-5	I	Interviewer evaluation of respondent comments indicating any difficulty in understanding harm caused by spills or the prevention program
	E-6	I	Interviewer evaluation of respondent having any difficulty understanding the voting question
8. Beliefs about the Scenario	C-1	D	Respondent's judgment about oil spill effects
	C-3	D	Respondent's judgment about effectiveness of prevention program
	C-4	D	Respondent's judgment about limit of special tax to single year
9. Ability/Willingness to Perform Task	E-7	I	Interviewer evaluation of whether respondent impatient to complete interview

Table 6.6. Cross-Tabulation Summary

Recommended Category	Source[a]	Choice Measure	Statistic of Association[b]	p-value[c]	Reject/ Not Reject Hypothesis of No Association
1. Income	D-12	**B1**	$\gamma = -0.006$	0.04	R
		B1CH	$\gamma = 0.024$	0.04	R
2. Prior Knowledge of Site	A-7	**B1**	$\chi^2_{(1)} = 8.366$	0.00	R
		B1CH	$\chi^2_{(4)} = 10.728$	0.00	R
	D-1	**B1**	$\chi^2_{(1)} = 4.811$	0.03	R
		B1CH	$\chi^2_{(1)} = 7.032$	0.01	R
3. Prior Interest in the Site	A-4	**B1**	$\chi^2_{(1)} = 1.942$	0.16	N
		B1CH	$\chi^2_{(1)} = 3.642$	0.06	N
	D-5	**B1**	$\chi^2_{(1)} = 5.746$	0.02	R
		B1CH	$\chi^2_{(1)} = 7.032$	0.01	R
	D-3	**B1**	$\chi^2_{(1)} = 7.573$	0.01	R
		B1CH	$\chi^2_{(1)} = 5.243$	0.02	R
	D-4	**B1**	$\chi^2_{(1)} = 9.878$	0.00	R
		B1CH	$\chi^2_{(1)} = 10.975$	0.00	R
4. Attitudes Toward Environment	A-1b	**B1**	$\chi^2_{(4)} = 47.561$	0.00	R
		B1CH	$\chi^2_{(4)} = 45.007$	0.00	R
	A-1e	**B1**	$\chi^2_{(4)} = 98.372$	0.00	R
		B1CH	$\chi^2_{(4)} = 100.051$	0.00	R
	A-2c	**B1**	$\chi^2_{(4)} = 123.309$	0.00	R
		B1CH	$\chi^2_{(4)} = 118.025$	0.00	R
	D-7	**B1**	$\chi^2_{(4)} = 44.130$	0.00	R
		B1CH	$\chi^2_{(4)} = 40.851$	0.00	R
5. Attitudes Toward Big Business	Oil companies should pay	**B1**	$\chi^2_{(1)} = 5.089$	0.02	R
		B1CH	$\chi^2_{(1)} = 6.295$	0.01	R
6. Distance to Site	Central Coast PSU's	**B1**	$\chi^2_{(1)} = 5.502$	0.02	R
		B1CH	$\chi^2_{(1)} = 6.655$	0.01	R
7. Understanding of Task	E-5	**B1**	$\chi^2_{(1)} = 2.780$	0.10	N
		B1CH	$\chi^2_{(1)} = 0.600$	0.44	N
	E-6	**B1**	$\chi^2_{(1)} = 4.009$	0.05	R
		B1CH	$\chi^2_{(1)} = 4.757$	0.03	R
8. Beliefs about Scenario	Oil Spill Effects: More Harm[d]	**B1**	$\chi^2_{(1)} = 46.032$	0.00	R
		B1CH	$\chi^2_{(1)} = 45.428$	0.00	R
	Oil Spill Effects: Less Harm[d]	**B1**	$\chi^2_{(1)} = 35.150$	0.00	R
		B1CH	$\chi^2_{(1)} = 30.477$	0.00	R
	Prevention Program: Might Work[e]	**B1**	$\chi^2_{(1)} = 35.333$	0.00	R
		B1CH	$\chi^2_{(1)} = 37.575$	0.00	R

(Continued over.)

Table 6.6. (Continued.)

Recommended Category	Source[(a)]	Choice Measure	Statistic of Association[(b)]	p-value[(c)]	Reject/ Not Reject Hypothesis of No Association
	Prevention Program: Not Work[(e)]	**B1**	$\chi^2_{(1)} = 70.391$	0.00	R
		B1CH	$\chi^2_{(1)} = 67.334$	0.00	R
	C-4	**B1**	$\chi^2_{(1)} = 13.183$	0.00	R
		B1CH	$\chi^2_{(1)} = 16.310$	0.00	R
9. Ability/Willingness to Perform Task	E-7	**B1**	$\chi^2_{(1)} = 4.173$	0.04	R
		B1CH	$\chi^2_{(1)} = 5.117$	0.02	R

(a) The source is the question number in the main survey unless otherwise indicated; see Table 6.5. Refused/not sure/not ascertained categories have been set to missing for the source variables and excluded from the cross-tabulations.

(b) When there are many categories, as with income, it is appropriate to report statistics such as the gamma and Kendall's tau-b. Here we report the gamma statistic, γ. For the other variables, we report the Pearson chi-squared statistic.

(c) The p-value is the probability level estimated for a Type-I error for a χ^2 statistic using a cross-tabulation of the choice measure and the recommended variable.

(d) Question C-1 was used to construct a (0,1) indicator, or dummy, variable for whether respondents felt oil spills off the Central Coast would cause "a lot more" harm (MOREHARM) or "a lot less" harm (LESSHARM) than that described in the survey.

(e) Question C-3 was used to construct a (0,1) indicator, or dummy, variable for whether respondents felt the prevention program would be "somewhat effective" (PMWORKS) or would be "not too effective" or "not effective at all" (PNOTWORK).

the choice *for* or *not-for* is not related to the variable. The reported p-value is the probability that the test result would call for incorrectly rejecting a *true* null hypothesis of no effect of the source variables on respondents' choices. The last column in the table reports the decision – assuming as a threshold the commonly used p-value of 0.05 – that would be made about differences in the distribution of responses between *for* and *not-for* choices (using both the **B1** and **B1CH** choice measures) and the distribution of responses in each of the category measures. The label "R" indicates that the null hypothesis of independence was rejected in favor of the alternative hypothesis of association between the choice measure and the variable; and "N" indicates that the null hypothesis was not rejected which suggests no association. For example, in the case of income, we reject the null hypothesis at the 95 percent confidence level that income does not affect the distribution of votes *for* and *not-for* since the calculated p-value of 0.04 is smaller than 0.05.

These cross-tabulations permit a simple test of association between respondents' choices and four different types of information.[23] The first type includes the characteristics and attitudes of respondents (*i.e.*, the first six categories in Table 6.6). Each of variables in these six categories displays a significant association at the 0.05 level with the **B1** and **B1CH** choice measures, with the exception of A-4 (shoreline visits) for which the relationship is suggestive ($p = 0.16$ for **B1** and $p = 0.06$ for **B1CH**). Variables with economic interpretations, such as income (D-12), as well as measures of activities that might be related to the injured resources, such as participation in various forms of saltwater recreation (D-3 and D-5) and identifying bird species (D-4), all influence the choice measure. The selection of the choice measure used in the cross-tabulation, **B1** or **B1CH**, does not influence this conclusion.

Respondents' environmental attitudes consistently relate to differences in decisions made about the program. Three variables from survey questions (A-1b, A-1e, and A-2c) about preferences for various environmental programs which were asked before the expected harm and prevention program questions were presented and a later question (D-7) which asks for a general self-evaluation on an environmentalist scale all significantly influence respondents' choices in the expected directions. For example, those respondents who feel that protecting coastal areas from oil spills on A-1e is very or extremely important and those who identify themselves as environmentalists on D-7 are both significantly more likely to vote *for* the program. We also indirectly measure respondents' attitudes towards big business by recording whether respondents volunteered that oil companies should pay all of the costs of the spill. The null hypothesis of no association is rejected. The cross-tabulation suggests that those who volunteered that oil companies should pay were significantly less likely to vote *for* the prevention program.

Variables in the *prior knowledge* and *distance to site* categories display the expected relationships with respondents' choices. Respondents indicating *prior knowledge*, *i.e.*, whether they were familiar with the five types of birds affected by past Central Coast spills and whether they had driven along the Central Coast on Highway 1, were more likely to vote *for* the program. *Distance to site*, measured here by a coastal proximity variable, also affected respondents' choices: respondents whose residences fall within the PSU's (see Chapter 4) along the coast between the San Francisco Bay and the greater Los Angeles area (CCOAST) were significantly more likely to vote *for*.

The second type of information summarized in Table 6.6 is respondents' *understanding of the task* as assessed by the interviewer. Those respondents having difficulty understanding either the harm caused by Central Coast oil spills or the prevention program (E-5) were not significantly more likely to

[23] The parameter estimates and z-statistics for the variables included in the construct validity model below (*see* Table 6.7) provide additional information about the extent to which each variable, controlling for the other variables in the equation, influences the percentage who voted *for* the program and in what direction this influence is exerted.

vote *for* the program; however, those having difficulty understanding the voting question (E-6; less than 2% of the total sample) were more likely to vote *for*.

The third type of information summarized in Table 6.6 is labeled "beliefs about scenario". As expected, respondents' perceptions of the effects of oil spills and of the effectiveness of the prevention program both affect the pattern of choices. For example, those respondents who thought that the effects of the oils spills were less severe than described and those who thought that the prevention program would not work were both significantly less likely to vote *for* the program. In addition, those who thought the tax would not be limited to just one year were less likely to vote *for*.

The fourth type of information summarized in the table is the respondent's *ability* or *willingness to perform the task*. The interviewers' evaluations of respondents' impatience to complete the interview offer an indirect gauge of respondent *willingness to perform the task*. The cross-tabulation suggests a significant association with impatient respondents tending to vote *not-for* the program.

The categories identified by the NOAA Panel may also be used as a basis for dividing the sample into subsamples; and separate WTP estimates may be computed for the various subsamples. For example, respondents who stated at D-1 that someone in their household had driven along the Central Coast on Highway 1 have a higher **B1CH** lower bound estimate than those who stated that someone had not (\$88.93 versus \$60.51; $t = 4.44$, $p < 0.001$). Also, the estimated mean derived from the choices of those respondents who indicated that oil spills cause less damage than described in the injury scenario is (as it should be) significantly smaller than that estimated from the choices of those who did not indicate that oil spills cause less damage; the difference between the subsample means is highly significant (\$41.58 versus \$92.51; $t = 7.40$; $p < 0.001$).

Very large differences between subsample means may be seen for questions relating to respondent preferences for the general class of resources protected by the program described in this study. For instance, the **B1CH** lower-bound estimate for those who said at A-1e that protecting coastal areas from oil spills was *extremely* important is \$121.57; while the lower bound on the sample mean for those who said it was *not important at all* is \$5.68 ($t = 2.91$; $p < 0.001$). These relationships reinforce the test results derived from the examination of the cross-tabulations recommended by the NOAA Panel.

6.6. Examination of Construct Validity Using a Multivariate Approach

The standard way to look at the simultaneous influence of multiple variables on respondent choices with respect to the oil spill prevention program is through the estimation of a multivariate choice function that relates respondent choices to various respondent characteristics. Such a function can also be used to demonstrate the *construct validity* of the CV results (Mitchell and Carson,

1989). Construct validity, one of the standard validity concepts widely accepted for use in evaluating models, refers to the degree to which a measure relates to other measures as predicted by economic theory, in this case, whether variation in the **B1CH** choice measure is systematically related to factors such as preferences for the object of choice, the cost of program, and the ability to pay for it.[24] Other factors that may be economically relevant include measures of respondents' evaluations of the expected harm and the characteristics of the prevention program. For example, respondents who thought oil spills would cause more harm than that described in the survey should be more likely to vote *for* the program.

6.6.1. *Definition and Interpretation of Covariates in Choice Function*

The dependent variable used in the choice model is the *vote for* indicator variable **B1CH**: respondents in favor of the tradeoff are coded 1 and other respondents are coded as 0. Economic theory suggests that B1AMT, the randomly assigned treatment variable, should have a negative coefficient and be a major determinant of respondents' choices (Deaton and Muellbauer, 1980), which is the observed result here. By itself in a simple probit equation, the coefficient on B1AMT is negative and highly significant with a z-statistic of -8.67 ($p < 0.001$).[25] Economic theory, however, only predicts that the percentage voting *for* should decline monotonically as the tax price increases. It does not suggest a specific functional form for this relationship. As a consequence, we allow for a flexible (*i.e.*, Box-Cox) transformation (Box-Cox, 1964; Greene, Greene and Seaks, 1995; Hanemann and Kanninen, 1999) on B1AMT in the multivariate choice function presented in Table 6.7.[26]

In the formulation presented, the coefficient on B1AMT and the Box-Cox λ parameter are highly correlated, masking their joint significance level.[27] A test of the joint significance of the two parameters can be performed using a likelihood ratio test. This test yields a $\chi^2(2) = 199.068$ which dictates rejection of the null hypothesis that $\beta \cdot \mathrm{B1AMT}^{\lambda}$ does not contribute to the model at $p < 0.001$.[28] Log-likelihood tests can also be used to compare the Box-Cox

[24] There is a long history of estimating construct validity equations in CV studies, see, *e.g.*, Knetsch and Davis (1966). For an example involving oil spill prevention, see Carson *et al.* (1992).

[25] With **B1CH** as the dependent indicator variable, the simple probit model yields a constant term of 0.3483 (0.0578) and a slope coefficient on B1AMT of -0.0044 (0.0005), where the standard errors are in parentheses.

[26] A Weibull choice model rather than the Box-Cox probit model used in Table 6.7 yields the same basic results; *see* Appendix H, Table 1.

[27] This high correlation between the Box-Cox parameter λ and the linear coefficient on a variable being scaled (B1AMT) has long been noted in the biometrics literature on fitting dose-response models (Morgan, 1992).

[28] Assuming λ was known *a priori* to be 0.3424, its estimated value in Table 6.7, the t-statistic on the transformed B1AMT would be -10.40 ($p < 0.001$).

Table 6.7. Multivariate Analysis of Construct Validity: Probit Estimates for **B1CH** Choice Valuation Function

Variable	Coding	Parameter Estimate	Z-Statistic	p-value (two-sided)	Variable Mean
CONSTANT	Equals 1 for all respondents	−2.0265	−2.88	0.004	NA
B1AMT	**B1** tax amount	−0.0801	−1.48	0.138	86.67
λ(B1AMT)	Box-Cox parameter	0.3424	1.51	0.132	NA
LINC1	Log of income if household income <$150,000; 0 otherwise	0.1945	3.19	0.001	9.00
LINC2	Log of income if household income $150,000; 0 otherwise	0.1573	2.65	0.008	0.44
NOTAX	Did not pay California taxes = 1; 0 otherwise	0.2147	1.25	0.212	0.10
CCOAST	Resides in Central Coast PSU (807, 812, 813, and 814) = 1; 0 otherwise	0.2187	2.31	0.021	0.45
COASTIP	A-1e protecting coastal areas very important or extremely important = 1; 0 otherwise	0.5031	4.01	0.000	0.78
WILDIP	A-2c spending to protect wildlife very important or extremely important = 1; 0 otherwise	0.5101	4.98	0.000	0.57
ENVIST	D-7 strong environmentalist or activist = 1; 0 otherwise	0.3717	3.09	0.002	0.21
LOWSPEND	Wants spending only on one or no programs (A-2a, A-2b, A-2d, and A-2e) = 1; 0 otherwise	−0.4835	−2.13	0.033	0.07
PAYVEH	D-16 prefer tax vehicle over higher prices or indifferent = 1; 0 otherwise	0.4993	5.00	0.000	0.40
HWY1	Traveled along the Central Coast on Highway 1 = 1; 0 otherwise	0.3328	2.35	0.019	0.88
FAMBIRD	Familiar with any of five types of birds often harmed in oil spills = 1; 0 otherwise	0.2700	1.94	0.053	0.86
MOREHARM	C-1 oil spills more harmful than described = 1; 0 otherwise	0.1867	1.74	0.083	0.35
LESSHARM	C-1 oil spills less harmful than described = 1; 0 otherwise	−0.3298	−2.30	0.021	0.16

(Continued over.)

Table 6.7. (Continued.)

Variable	Coding	Parameter Estimate	Z-Statistic	p-value (two-sided)	Variable Mean
PMWORKS	C-3 expect program to be somewhat effective = 1; 0 otherwise	−0.6403	−6.46	0.000	0.39
PNOTWORK	C-3 expect program to be not too effective or not effective at all = 1; 0 otherwise	−1.5289	−7.18	0.000	0.08
PAYMORE	C-4 does not think will only have to pay special tax for one year = 1; 0 otherwise	−0.2425	−2.56	0.011	0.44
PROTEST	Stated oil companies should pay for program or that oil companies would pass program costs on to consumers = 1; 0 otherwise	−0.7283	−5.93	0.000	0.20
	N = 1085				
	Log(L) = −500.26				
	Pseudo $R^2 = 0.335$				

formulation against the $\lambda = 1$ linear specification and the log specification: both of these simpler specifications are rejected in favor of the Box-Cox model ($\chi^2(1) = 10.678$ ($p = 0.001$), and $\chi^2(1) = 4.462$ ($p = 0.035$), respectively).

The other variables selected for inclusion in the choice model can be grouped into five broad categories: economic and demographic characteristics, preferences and attitudes, interest in and use of the affected natural resources, evaluations of the expected harm and prevention program, and interpretations of the payment mechanism.[29] The model in Table 6.7 explains a substantial fraction of the variability in the choices. Respondents with some sets of characteristics are predicted by the estimated choice model to be willing to pay less than one dollar; while respondents with other sets of characteristics are predicted to be willing to pay as much as several hundred dollars.

The economic and demographic variables are LINC1, LINC2, NOTAX, and CCOAST.[30] The model in Table 6.7 allows income to have a different coefficient

[29] Note that Table 6.7 reports p-values for two-sided hypothesis tests. In most instances, the hypothesis about the coefficient on a particular test is of the one-sided form (*e.g.*, a null hypothesis that respondents who do not think the program works are as likely to vote *for* the program as other respondents versus the alternative that they are less likely). For one-sided hypothesis tests, the reported (two-sided) p-values should be divided by 2.

[30] Missing values for income (n = 86) have been replaced with an estimate based on the median income in the 1993 zip code, housing type, education, gender, race, age, and qualitative variables for the number of employed adults in the household. Tables 2 and 3 in Appendix H present more detailed definitions of the variables included in the income prediction equation and the model for estimating income, respectively. Excluding from the sample the households who did

depending upon the level of household income.[31] Two income classes are identified, those below \$150,000 (LINC1) and those of \$150,000 and above (LINC2). The coefficients of both income terms are positive and statistically significant with p-values less than 0.01.

NOTAX is an indicator variable for households that reported they did not pay California income taxes in 1994. The coefficient on this variable is positive but not significant at conventional significance levels. However, the one-sided p-value, 0.106, is suggestive that those not paying taxes are willing to pay more than their characteristics would otherwise imply.

The demographic variable in the model is CCOAST, a qualitative variable identifying the respondent's location in relationship to the area of the natural resource injuries. Respondents living in PSU's between San Francisco Bay and the greater Los Angeles area (CCOAST = 1) are significantly more likely to vote *for* the program than those in the rest of the state.

The preference and attitude variables are COASTIP, WILDIP, ENVIST, LOWSPEND, and PAYVEH. The three preference variables COASTIP, WILDIP, and ENVIST are directly related to the environment. COASTIP is a qualitative variable identifying those respondents who, at question A-1e, rated preventing oil spills in coastal areas as *very important* or *extremely important*. WILDIP, also a qualitative variable, identifies those respondents who, at A-2c, thought spending to protect wildlife is *very important* or *extremely important*. ENVIST identifies individuals who, at D-7, considered themselves to be either strong environmentalists or environmental activists. The positive signs on the COASTIP, WILDIP, and ENVIST coefficients and the associated p-values suggest that those who support the relevant class of environmental programs and who identify themselves as environmentalists are significantly more likely to vote for the program.

The two attitude variables, LOWSPEND and PAYVEH, relate more generally to respondents' attitudes about government programs. LOWSPEND identifies those respondents who view spending as *not too important* or *not important at all* for at least three of four other programs asked about in question A-2 (*i.e.*, job training for the unemployed, shelters for the homeless, lifeguards

not report income does not change the sign or significance of the income measure or the role of any other variables; *see* Table 4 in Appendix H. It does reduce the sample from 1085 to 999, so the p-values for some of the tests for relationships between these variables and respondents' choices necessarily decrease somewhat.

31 If one allows the income coefficient to vary with the level of household income, the effect of log income on the probability of favoring the program is still positive and significant with a p-value of 0.006. This result holds regardless of the treatment of missing values for income. The one-variable income specification can be rejected in favor of the two-variable specification used here using a likelihood-ratio test $\chi^2(11) = 2.86$, $p = 0.09$). If LINC1 in turn is split into two categories, one consisting of those households with income greater than the median California household income and one consisting of those below, the estimated income effect in the second category (LINC2) is still smaller but not significantly so.

at state beaches, and public transportation in Los Angeles).[32] PAYVEH is defined from respondents' evaluations of whether taxes are the appropriate way to pay for new programs to protect the environment. Both LOWSPEND and PAYVEH have the expected signs and are significant: those not favoring increased government spending are less likely to vote for the program and those favoring the use of government taxes to effect environmental improvements or who are indifferent between higher taxes and higher prices to effect environmental improvements are more likely to vote for the program.

The next two variables reflect interest and use of the affected natural resources: driving along the Central Coast on Highway 1 (HWY1) and familiarity with at least one of the five species of birds most often harmed by past oil spills (FAMBIRD). HWY1 and FAMBIRD are both positive and significant: individuals whose activities and knowledge are related to the Central Coast (HWY1) and the five species of birds (FAMBIRD) are more likely to vote for the program.

The next four variables in Table 6.7 are related to respondents' evaluations of the expected harm and the prevention program. Those respondents who thought that oil spills along the Central Coast over the next 10 years would cause more harm than that described in the survey (MOREHARM) should be and were more likely to vote *for* the program; and those who thought that oil spills would cause less harm (LESSHARM) should be and were more likely to vote *not-for* the program. Their respective coefficients in Table 6.7 have the expected sign, and both effects are statistically significant. The effects of MOREHARM and LESSHARM on willingness to pay offset almost exactly.[33] Also, those who thought the program would be somewhat effective (PMWORKS) or not effective (PNOTWORK) should be and were less likely to vote *for* the program. Again the effects are highly significant. Had all respondents thought that the program would be completely effective, the percentage of *for* votes would have been higher. These results on the perceived degree of harm prevented and the perceived program effectiveness provide strong within-sample evidence that respondents are sensitive to the scope of the good they were asked to value.

The final two variables relate to the respondents' interpretation of the payment mechanism. PAYMORE identifies those respondents who thought the tax payment might not be limited to just one year. PROTEST identifies those respondents who protested at any point during the interview that either the oil companies should pay for all of the program costs or that the oil companies would pass on their share of the costs to consumers in the form of higher gas

32 Note that the other program asked about in the A-2 series, spending on prisons, was not included in LOWSPEND as its inclusion resulted in perfect failure (*i.e.*, all of the respondents who meet this more inclusive criterion voted *not for* the program).

33 The absolute value of the coefficient on LESSHARM is almost twice that of MOREHARM. However, the percentage of respondents giving a MOREHARM answer is more than double that of those giving a LESSHARM answer.

and oil prices. Those who did not think they would have to pay the amount for only one year (PAYMORE) should be and were less likely to vote *for* the program as were those who protested the payment mechanism (PROTEST).

Other information on demographics, knowledge, and attitudes or behaviors were also considered in our evaluation of the construct validity model reported in Table 6.7. The two demographic variables, age and gender, were not significant determinants of choices when income, attitude, and program-evaluation variables were included in the model. However, in bivariate relationships with **B1CH**, age (negative relationship) and gender (positive relationship with female) were statistically significant. Other variables, positive and significant in bivariate relationships with **B1CH**, but not significant in the model in Table 6.7, include four variables related to the use of and interest in natural resources: saltwater boating or fishing (D-3), bird-watching (D-4), going to the beach (D-5), and watching television programs about animals and birds in the wild (D-6).

6.6.2. *A Cluster Analysis Interpretation*

The multivariate regression results presented in Table 6.7 is the standard approach in the contingent valuation literature to demonstrating construct validity (Mitchell and Carson, 1989). That equation parsimoniously summarizes the dependence of the probability of a *vote for* the program on the tax amount and various other covariates. Cluster analysis, an alternative approach used particularly in marketing research, partitions the sample into clusters of respondents based on the respondents' values for the covariates.[34] One can then look at the probability of a respondent voting *for* the program given the particular cluster to which the respondent has been assigned.

A large number of clustering algorithms have been proposed (Hartigan, 1975; Kaufman and Rousseeuw, 1990). Perhaps the most commonly used clustering approach is a partitioning method known as k-means clustering. Given a predetermined number of clusters, k, the sample is divided into k distinct clusters which minimize the within cluster variation through the choice of the multidimensional centers of each of the k clusters and assigning particular observations to particular clusters. Each of the k cluster centers can be thought of as a representative respondent. The variables are typically normalized so that each variable plays an equal role in terms of the variance function being

[34] The usual objective of cluster analysis in marketing research is to segment potential consumers into groups that differ by one or more key characteristics (Lunn, 1986). These segments are often based on demographic variables, differences in purchasing patterns, or the marketing channel used.

Table 6.8. Cluster Centers

Variable	Cluster A	Cluster B	Cluster C	Cluster D
COASTIP	0.406	0.141	−0.436	−0.812
WILDIP	0.757	0.173	−0.931	−0.721
FAMBIRD	0.193	−0.210	−0.228	0.150
HWY1	0.077	−0.524	−0.012	0.275
ENVIST	0.241	0.141	−0.325	−0.198
CCOAST	0.051	0.109	−0.043	−0.303
MOREHARM	0.302	0.130	−0.347	−0.519
LESSHARM	−0.204	−0.169	0.269	0.276
PMWORKS	−0.097	−0.091	0.145	0.056
PNOTWORK	−0.212	−0.096	0.170	0.779
PAYMORE	−0.039	0.054	0.026	0.064
PAYVEH	0.220	0.223	−0.280	−0.417
LOWSPEND	−0.268	−0.154	−0.268	3.722
PROTEST	−0.083	−0.182	0.082	0.441
LINC	0.169	−1.783	0.206	0.271
NOTAX	−0.332	2.879	−0.332	0.002

minimized.[35] The covariate values for each of the representative respondents are expressed as deviations from the average respondent in the sample as a whole.

The *k*-means cluster analysis is performed on the set of predictor variables used in the model presented in Table 6.7 with the exception of CONSTANT which does not vary across respondents and B1AMT which was randomly assigned to respondents.[36] The value of *k* chosen for this analysis is four.[37] With respect to the fit of the cluster solution, the sum of the within cluster variance is 88.5% of the sum of the variance of the sample; the pseudo F-statistic is 100.91; and the over-all R^2 measure is 0.21.

The four representative respondents, *i.e.*, the centers of the four clusters, are given in Table 6.8. Each cluster center takes on a value for each of the 16 predictor variables used in the cluster analysis. Positive values represent a

[35] Normalization of a variable is accomplished by subtracting the variable's mean value from each observation in the data set and then dividing by the variable's standard deviation. The normalized variable has a mean value of zero and a standard deviation of one. The value of a normalized variable is interpreted in terms of the number of standard deviations from the mean.

[36] LINC1 and LINC2 have been added together to form LINC, the log of the respondent's household income. There is no gain in the cluster approach to using two income variables as the clustering algorithm can perform its partitioning at any point along the income distribution.

[37] The choice of *k* in a cluster analysis is largely dependent upon the purpose for which the analysis is intended and the nature of the data being clustered. Allowing too few clusters can suppress key detail in the data. Allowing too many clusters makes interpretation difficult and eventually will largely reproduce the regression results already provided in Table 6.7. We have chosen *k* equal to four as a compromise. Much of the same insight is gained *if* *k* is equal to three or *k* is equal to five.

cluster center for that variable which is larger than the average of that variable for the entire sample, and negative values represent the opposite. For instance, looking at the row for the COASTIP variable, the representative respondent for Cluster A was more likely than the average respondent in the sample to think that protecting coastal areas was important. The representative respondent for Cluster B was slightly more likely than the average respondent to think that protecting coastal areas was important; the representative respondent for Cluster C was less likely than the average respondent; and the representative respondent for Cluster D was much less likely. Such a comparison of the values of the normalized variables across the four clusters can be made for each variable. The *k*-means clustering algorithm assigns 521 respondents to Cluster A, 105 respondents to Cluster B, 389 respondents to Cluster C, and 70 respondents to Cluster D.

The first column of Table 6.8, which contains cluster A's variable centers for each variable, presents a very clear impression of the type of respondent assigned to this cluster. They are more likely than the average respondent to believe that protecting coastal areas (COASTIP) and wildlife (WILDIP) is important and to identify themselves as strong environmentalists (ENVIST). They are somewhat more familiar with the specific resources (FAMBIRD, HWY1). They are also are more likely than the average respondent to believe that there would be more injury from oil spills (MOREHARM, LESSHARM) and more likely to think that the plan would be effective (PMWORKS, PNOTWORK). They are also more likely to prefer the tax payment vehicle (PAYVEH) and less likely to be unsupportive of other public programs (LOWSPEND) than the average respondent in the sample. They have higher incomes (LINC) than the average respondent in the sample and are less likely than the average respondent not to pay California income taxes (NOTAX). We can look at the voting implications of being assigned to this cluster in two ways. One way is to compare the percentage of respondents voting *for* in this cluster, 68.1%, compared to the 48.8% voting *for* in the sample as a whole, or the 31.0% voting *for* in the sample excluding Cluster A respondents.[38] The other way is to look at the Turnbull lower bound on the sample mean. Table 6.9 summarizes the information on a probability of a *for* vote and the Turnbull lower bounds for the sample as a whole and each of the four clusters. Cluster A respondents have a Turnbull lower bound mean of $121.53, over 40% higher than the sample as a whole ($85.39). Cluster A's respondents comprise just under half the sample but over two-thirds ($57.20) of the total magnitude of the sample Turnbull lower bound mean ($85.39).

Looking down the column for Cluster B in Table 6.8, a picture emerges of a set of respondents who have moderately stronger environmental preferences

[38] Because the B1AMT's were assigned to respondents independently of their characteristics, and B1AMT was not used in determining the clusters, it is meaningful to look at the average probability of a *for* vote across the different clusters.

Table 6.9. Summary Statistics for Sample and Cluster

Cluster	Sample Size N	Percentage Voting For	Turnbull Lower Bound Mean	Standard Error of the Mean
Sample	1085	48.8	85.39	3.90
A	521	68.1	121.53	5.91
B	105	46.7	81.39	11.63
C	389	30.0	50.18	5.58
D	70	12.9	20.77	24.81

(COASTIP, WILDIP, ENVIST) than the average respondent in the sample but who have less familiarity with the resource depicted in the scenario (FAMBIRD, HWY1). The dominant characteristic of Cluster B respondents, however, is their low income (LINC) and their much higher likelihood of not having paid California income taxes (NOTAX).[39] Cluster B respondents have a 46.7% probability of providing a *for* vote and have a Turnbull lower bound mean of $81.39. Both of these are close to the sample average probability.

Looking down the column for Cluster C, the picture that emerges is that of a group of respondents who should have a lower probability of voting for the program. The representative respondent of this cluster has less preference for environmental amenities (*i.e.*, COASTIP, WILDIP, ENVIST), is less familiar with the specific resources (FAMBIRD, HWY1), is more likely to think that the harm from a spill will be less than that described by the scenario (MOREHARM, LESSHARM), is more skeptical about the plan working (PMWORKS, PNOTWORK), and dislikes the payment vehicle (PAYVEH). The only positive factor with respect to a *for* vote is a higher income (LINC). Cluster C is the second largest cluster with 389 respondents. These respondents have a 30.0% probability of having a *for* vote and have a Turnbull lower bound mean of $50.18.

Looking down the column for Cluster D, one sees a representative respondent who is fairly negative with respect to the environment (COASTIP, WILDIP, ENVIST). This respondent representative is more familiar than the average respondent with the specific resources (FAMBIRD, HWY1) but less likely to live anywhere along the coast between the greater Los Angeles area and the San Francisco Bay area (CCOAST). This respondent is more likely to be skeptical about the amount of harm done by oil spills (MOREHARM, LESSHARM) and believes that the proposed program is not likely to work (PNOTWORK). The representative respondent of this cluster dislikes the payment vehicle; but the most distinguishing characteristics of Cluster D respondents is their dislike of spending on any government program (LOWSPEND) and their much greater proclivity to protest that the oil companies should be paying (PROTEST). Cluster D is the smallest cluster with 70

[39] The influence of respondents not paying California income taxes is considered in section 6.8.

Table 6.10. Box-Cox Model Using Cluster Indicators

Variable	Parameter Estimate	Z-Statistic	p-value (two-sided)	Variable Mean
CONSTANT	1.1716	6.85	0.000	NA
B1AMT	−0.0526	−1.21	0.226	86.67
λ(B1AMT)	0.4628	2.37	0.021	NA
CLUSTER B	−0.5245	−3.73	0.000	0.10
CLUSTER C	−1.0582	−11.67	0.000	0.36
CLUSTER D	−1.6827	−8.27	0.000	0.06

N = 1085
Log(L) = −616.86
Pseudo $R^2 = 0.180$

respondents. Given the above characteristics of the representative respondent of this cluster one would expect a low probability of a vote *for*; indeed, Cluster D respondents have a 12.9% probability of voting *for* the program; and they have a Turnbull lower bound mean of $20.77. As one might expect, none of the Cluster D respondents are willing to pay the highest amount asked, $220; while over 65% of the respondents voting *for* this dollar amount come from the Cluster A respondents.

An equation similar to that in Table 6.7 may be created by substituting indicator variables of cluster membership for the predictor variables other than B1AMT. This probit model is reported in Table 6.10. The Cluster A indicator variable has been absorbed into the CONSTANT term. Membership in Cluster B, Cluster C, or Cluster D decreases the probability of a *for* vote. As one would expect from the information in Table 6.9, the parameter estimates become more negative as one goes from Cluster B to C to D. Note that the parameter estimates on the cluster membership indicator variables are all highly significant.[40] The pseudo R-square of this model is a little over half that of the much larger model reported in Table 6.7, reflecting the fact that this set of four clusters has substantial explanatory power but does not use all the relevant information contained in the model presented in Table 6.7.

6.7. Sensitivity of WTP Estimate

In this section, using the construct validity equation reported in Table 6.7, we examine the effects of respondent assumptions that deviate in some way from

[40] The basic set of Box-Cox parameters are, as in Table 6.7, highly correlated with each other; this correlation is responsible for their fairly small overall z-statistics. In a probit model of the cluster indicators with either B1AMT or log(B1AMT) as the stimulus variable, the z-statistic on the stimulus variable is over 9 ($p < 0.001$). The linear form of the model in Table 6.10 can be rejected using a likelihood ratio test at $p = 0.02$ and the log form of the model at $p = 0.01$.

the scenario presented in the survey instrument. The information about these assumptions comes from respondents' answers to the Section C questions about what they had in mind when they voted for or against the plan and from respondents' answers to questions concerning the payment mechanism. We also looked at the sensitivity of the WTP estimate to possible protest responses identified by the PROTEST variable used in Table 6.7.

One may look at a number of different summary statistics in regard to shifts in the WTP distribution related to these deviations. In this section, we focus on the shift in the estimate of median WTP, \$60.56,[41] because this statistic is fairly robust to statistical assumptions about the general shape of the underlying WTP distribution.[42]

The first possible shift we consider is related to those respondents who believed that the harm would be more or less than the harm described. One of two dummy variables, either MOREHARM or LESSHARM, in the construct validity equation in Table 6.7 has a value set to one to represent the particular deviation from the desired perception. Setting the value of these two dummy variables to zero effectively forces the perceptions to the *same* harm category. This adjustment, however, does not change the estimate of median household willingness to pay by more than a few cents. As noted above, the combined MOREHARM and LESSHARM effect is almost exactly off-setting.

Another possible shift relates to the perceived effectiveness of the program. Ideally, all respondents would have perceived the plan as being effective. One of two dummy variables, PMWORKS and PNOTWORK, in the construct validity equation has a value of one if a respondent indicated belief that the plan would not be completely or mostly effective. Setting both of these dummy variables to zero effectively forces the perception that the plan is effective. This adjustment increases the estimated median willingness to pay by \$58.86 to \$119.42. This increase is a dramatic indication of the importance of belief in the program's effectiveness.

41 The median is the point in the WTP distribution above which 50 percent of the respondents are predicted to be willing to pay more and below which 50 percent are willing to pay less.

42 In contrast, the estimate of mean WTP is quite sensitive to distributional assumptions and tends to be dominated by the assumption that is made regarding a very small percentage of observations in the right tail of the distribution. An additional issue with using the Box-Cox model in Table 6.7 (Collins, 1991) is that there are a number of technical difficulties associated with inverting that model to get mean estimates, particularly for individual observations with predicted estimates close to zero. If one sets to zero the predicted values that are close to zero, the changes in the predicted Box-Cox means are generally similar in magnitude to the changes in the predicted medians reported in this section. The Turnbull estimate of the lower bound on the sample mean used in earlier sections of this chapter avoids all of these difficulties. The Turnbull framework, does however, have the disadvantage (relative to the Box-Cox) that it is neither computationally or straightforward to generalize that framework to look at the implications of changes in particular covariate values while holding other covariate values constant. One can, of course, use the Turnbull framework to look at the differences in WTP based on any rule that divides observations into a small number of finite groups as presented earlier.

A third possible shift is suggested by some respondents who believed that they would have to pay the special tax for more than one year. Ideally all respondents would have perceived the tax as limited to a one-time payment. Setting the dummy variable PAYMORE to zero effectively forces the perception that the tax was a one-time payment. This adjustment increases the estimated median WTP by $18.25 to $78.81.

A fourth possible shift is that related to protest responses. As noted in Table 6.7, PROTEST identifies those respondents who protested at any point during the interview that either the oil companies should pay for all of the program costs or that the oil companies would pass on their share of the costs to consumers in the form of higher gas and oil prices. Setting the PROTEST dummy variable to zero in the construct validity model forces out that consideration and increases the estimated median WTP by $23.77 to $84.33.

In sum, these shifts in the median WTP suggest that the overall effect of respondent assumptions that deviate from these four scenario features is to bias downward the estimated willingness to pay for the program put forth in the survey scenario. Although the direction of the estimated changes in median WTP are likely to be a reliable indicator of the direction of the bias in the Turnbull estimate reported earlier, an extremely strong set of assumptions would be necessary to justify translating the absolute magnitude of these changes into specific changes in the Turnbull estimate of the lower bound on the sample mean WTP.

6.8. Correction for Non-Taxpayers

The payment vehicle used in this study – a one-time increase in California income taxes – presents a problem different in type from that of the potential divergences between respondent perceptions and the scenario presented: respondents not currently paying state income taxes do not necessarily treat the tax payment obligation in the same way as those respondents paying state income taxes. The sign and significance of NOTAX in the multivariate choice model suggests that this group of respondents is willing to pay more than other respondents with otherwise identical characteristics. Below, we treat the **B1CH** responses of those respondents who did not pay California income taxes in 1994 as *not-for* responses. By recoding from *for* to *not-for* the votes of the 108 respondents who did not pay California income taxes in 1994 and who voted *for* the program, we have made the most conservative possible adjustment as it effectively sets the lower-bound Turnbull estimate for this group of respondents to zero.[43] This choice measure is referred to as **B1CHNT** below.

Table 6.11 reports the distribution of **B1CHNT** by B1AMT and Table 6.12 reports the Turnbull estimate of the lower bound on the sample mean for this

[43] *See* Appendix F.

Table 6.11. **B1CHNT** Choice Measure by B1AMT

B1AMT	For	Not For
$5	62.10%	37.90%
$25	51.85%	48.15%
$65	45.23%	54.77%
$120	35.36%	64.64%
$220	25.44%	74.56%
	$\chi^2_{(4)} = 71.98$; $p < 0.001$	

choice measure. As for the **B1** and **B1CH** choice measures, a $\chi^2(4)$ test (71.98) for the **B1CHNT** measure rejects the hypothesis ($p < 0.001$) that responses are not sensitive to B1AMT. As shown in Table 6.12, the estimated lower bound on the sample mean for the **B1CHNT** choice measure is $76.45[44] with a standard error of $3.78.[45] This estimate, smaller than that from the **B1CH** choice measure ($85.39, with a standard error of $3.90), represents a conservative adjustment to the lower-bound estimate on the sample mean WTP.

6.9. Summary

In this chapter, after adjusting for non-taxpayers, we estimate a conservative lower bound on the average *ex ante* economic value of the oil spill prevention program to sample households of $76.45, with a standard error of $3.78. Two

Table 6.12. Turnbull Estimate of WTP Distribution and Lower Bound on the Sample Mean: **B1CHNT** Choice Measure [N = 1,085]

Lower Bound of Interval	Upper Bound of Interval	Probability of Voting *For* at Upper Bound	Change in Density
$0	$5	0.6210	0.3790
$5	$25	0.5185	0.1025
$25	$65	0.4523	0.0662
$65	$120	0.3536	0.0987
$120	$220	0.2544	0.0992
$220	∞	0.0000	0.2544

Log-Likelihood −707.73
Estimate of lower bound on sample mean $76.45
Standard error of the estimate $3.78

44 For the **B1CHNT** choice measure, the Turnbull estimate of the lower bound on the sample mean using the sample weights (Appendix B.10) is $77.36, $0.91 higher than the unweighted estimate. The standard error of the weighted estimate is $3.73.

45 The z-statistics for the five change-in-density parameters estimated by the model are 11.56, 2.17, 1.42, 2.06, and 2.17, respectively.

types of quantitative evaluations provide extensive quantitative evidence on the validity and reliability of the choice data collected in this survey, complementing the analysis presented in the previous chapter.

The first such evaluation consists principally of bivariate relationships in the form of cross-tabulations between the information variables recommended by the NOAA Panel and the **B1** and **B1CH** choice measures. Overall, the bivariate analysis provides support for the presence of relationships that economic theory suggests should influence respondents' choices regarding the prevention program.

The second evaluation considers the same issues in a more structured format using a conventional multivariate choice model. A probit model is estimated to identify the determinants of **B1CH** choices. In each instance, the factors hypothesized to be associated with the choices are found to be consistent with prior expectations; and the relationships are statistically significant determinants of **B1CH**. Moreover, these effects are robust and generally do not change much with the specific coding of the variables involved. This construct validity equation is used to examine the effect that various adjustments would have on the WTP estimated by the model. In adjusting the WTP estimate to account for shifts due to perceptions of harm differing from those presented in the survey, the increase in WTP due to those perceiving more harm is almost exactly offset by the decrease in WTP due to those perceiving less harm. Adjustments for perceptions that the program would be less than effective, perceptions that the tax would not be one-time, and for protest responses, all result in substantially higher estimates.

References

AAPOR (2000), 'Standard Defintions: Final Dispositions of Case Codes and Outcomes for Survys', Ann Arbor, MI: American Association of Public Opinion Research.

Adamowicz, W., P. Boxall, M. Williams, and J. Louviere (1998), 'Stated Preference Approaches for Measuring Passive Use Values Choice Experiments and Contingent Valuation', *American Journal of Agricultural Economics* **80**, 64–76.

Adamowicz, W., J. Louviere and M. Williams (1994), 'Combining Revealed and Stated Preference Methods for Valuing Environmental Amenities', *Journal of Environmental Economics and Management* **26**, 271–292.

Alberini, A., B. Kanninen, and R.T. Carson (1997), 'Modeling Response Incentive Effects in Dichotomous Choice Contingent Valuation Data', *Land Economics* **73**, 309–324.

Andreoni, J. (1989), 'Giving with Impure Altruism – Applications to Charity and Ricardian Equivalence', *Journal of Political Economy* **97**, 1447–1458.

Arndt, J. and E. Crane, (1975) 'Response Bias, Yea-Saying, and the Double Negative', *Journal of Marketing Research* **12**, 218–220.

Arrow, K., R. Solow, P.R. Portney, E.E. Leamer, R. Radner, and H. Schuman (1993), 'Report of the NOAA Panel on Contingent Valuation', *Federal Register* **58**, 4601–4614.

Atkinson, A.B. and J.E. Stiglitz (1980), *Lectures on Public Economics.* New York: McGraw-Hill.

Audience Research, Inc. (1958), *A Study of Outdoor Recreational Activities and Preferences of the Population Living in the Region of the Delaware River Basin.* Report prepared for the National Park Service.

Ayer, M., H.D. Brunk, G.M. Ewing, W.T. Reid, and E. Silverman (1955), 'An Empirical Distribution Function for Sampling with Incomplete Information', *Annals of Mathematical Statistics* **26**, 641–647.

Baron, D.P. (1994), 'Electoral Competition With Informed And Uninformed Voters', *American Political Science Review* **88**, 33–47.

Bateman, I., R.T. Carson, Brett Day, W.M. Hanemann, N. Hanley, T. Hett, M. Jones-Lee, G. Loomes, S. Mourato, E. Özdemiroglu, D. Pearce R. Sugden, and J. Swanson (2002), *Economic Valuation with Stated Preference Techniques: A Manual.* Northhampton, MA: Edward Elgar.

Bergstrom, T.C., D.L. Rubinfeld, and P. Shapiro (1982), 'Micro-Based Estimates of Demand Functions for Local School Expenditures', *Econometrica* **50**, 1183–1205.

Bishop, R.C. and T.A. Heberlein (1979), 'Measuring Values of Extra-Market Goods: Are Indirect Measures Biased?', *American Journal of Agricultural Economics* **61**, 926–930.

Black, D. (1958), *The Theory of Elections and Committees.* Cambridge: Cambridge University Press.

Bohm, P. (1994), 'CVM Spells Responses to Hypothetical Questions', *Natural Resources Journal* **34**, 37–50.

Boman, M.G. Bostedt and B. Kristrom (1999), 'Obtaining Welfare Bounds in Discrete-Response Valuation Studies: A Non-parametric Approach', *Land Economics* **75**, 284–294.

Box, G.E.P. and D.R. Cox (1964), 'An Analysis of Transformation', *Journal of Royal Statistical Society Series B* **26**, 211–252.

Braden, J.B., C.D. Kolstad, and D. Miltz (1991), 'Introduction', in J. Braden and Charles

Kolstad, eds., *Measuring the Demand for Environmental Commodities.* Amsterdam: North-Holland.

Brown, T.C., P.A. Champ, R.C. Bishop, and D.W. McCollum (1996), 'Which Response Format Reveals the Truth About Donations to a Public Good', *Land Economics* **72**, 152–166.

Cameron, T.A. and M.D. James (1987), 'Efficient Estimation Methods for "Closed-Ended" Contingent Valuation Surveys', *Review of Economics and Statistics* **69**, 269–276.

Carson, R.T. (1991), 'Constructed Markets', in J. Braden and Charles Kolstad, eds., *Measuring the Demand for Environmental Commodities.* Amsterdam: North-Holland.

Carson, R.T. (1997a), 'Contingent Valuation and Tests of Insensitivity to Scope', in R.J. Kopp, W. Pommerhene, and N. Schwartz, eds., *Determining the Value of Non-Marketed Goods: Economic, Psychological, and Policy Relevant Aspects of Contingent Valuation Methods*, Boston: Kluwer.

Carson, R.T. (1997b), 'Contingent Valuation: Theoretical Advances and Empirical Tests Since the NOAA Panel', *American Journal of Agricultural Economics* **79**, 1501–1507.

Carson, R.T. (2000), 'Contingent Valuation: A User's Guide,' *Environmental Science and Technology* **34**, 1413–1418.

Carson, R.T. (forthcoming), *Contingent Valuation: A Comprehensive Bibliography and History.* Northhampton, MA: Edward Elgar.

Carson, R.T., M.B. Conaway, W.M. Hanemann, J.A. Krosnick, K.M. Martin, D.R. McCubbin, R.C. Mitchell, and S. Presser (1996), *The Value of Preventing Oils Spills to Natural Resources Along Calfornia's Central Coast*, 2 vols., La Jolla, CA: Natural Resource Damage Assessment, Inc.

Carson, R.T. and N.E. Flores (1996), 'Another Look at "Does Contingent Valuation Measure Preferences?: Experimental Evidence" How Compelling is the Evidence?', Discussion Paper 95–21, Department of Economics, University of California, San Diego.

Carson, R.T., N.E. Flores, and N.M. Meade (2001), 'Contingent Valuation: Controversies and Evidence,' *Environmental and Resource Economics* **19**, 173–210.

Carson, R.T., N.E. Flores and W.M. Hanemann (1998), 'Sequencing and Valuing Public Goods', *Journal of Environmental Economics and Management* **36**, 314–323.

Carson, R.T., N.E. Flores, K.M. Martin and J.L. Wright (1996), 'Contingent Valuation and Revealed Preference Methodologies: Comparing the Estimates for Quasi-Public Goods', *Land Economics* **72**, 80–99.

Carson, R.T., N.E. Flores and R.C. Mitchell (1999), 'The Theory and Measurement of Passive Use Value', in I.J. Bateman and K.G. Willis, eds., *Valuing Environmental Preferences: Theory and Practice of the Contingent Valuation Method in the US, EC, and Developing Countries* (Oxford: Oxford University Press).

Carson R.T., T. Groves, and M. Machina (1999), 'Incentive and Informational Properties of Preference Questions,' Plenary Address, European Association of Environmental and Resource Economists, Oslo, Norway, June 1999.

Carson, R.T., W.M. Hanemann, R.J. Kopp, J.A. Krosnick, R.C. Mitchell, S. Presser, P.A. Ruud, and V.K. Smith (1994), *Prospective Interim Lost Use Value Due to DDT and PCB Contamination in the Southern California Bight: Volumes 1 & 2.* Report to the National Oceanic and Atmospheric Administration. La Jolla, CA: Natural Resource Damage Assessment, Inc.

Carson, R.T., W.M. Hanemann, R.J. Kopp, J.A. Krosnick, R.C. Mitchell, S. Presser,

P.A. Ruud, V. K. Smith, M.B. Conaway, and K.M. Martin (1997), 'Temporal Reliability of Estimates from Contingent Valuation', *Land Economics* **73**, 151–163.

Carson, R.T., W.M. Hanemann., R.J. Kopp, J.A. Krosnick, R.C. Mitchell, S. Presser, P.A. Ruud, V. K. Smith, M.B. Conaway, and K.M. Martin (1998), 'Referendum Design and Contingent Valuation: The NOAA Panel's No-Vote Recommendation', *Review of Economics and Statistics* **80**, 484–487.

Carson, R.T., W.M. Hanemann and R.C. Mitchell (1987), 'The Use of Simulated Political Markets to Value Public Goods', Discussion Paper 87–7. Department of Economics, University of California-San Diego.

Carson, R.T., M.W. Hanemann and D. Steinberg (1990), 'A Discrete Choice Contingent Valuation Estimate of the Value of Kenai King Salmon', *Journal of Behavioral Economics* **19**, 53–68.

Carson R.T. and Y. Jeon (2000), 'On Overcoming Informational Deficiencies in Estimating Willingness-To-Pay Distributions', Paper presented at the American Agricultural Economics Association meeting, Tampa, Florida, June 2000.

Carson, R.T., J. Louviere, D. Anderson, P. Arabie, D. Bunch, D. Hensher, R.M. Johnson, W.F. Kuhfeld, D. Steinberg, J. Swait, H. Timmermans, J.B. Wiley (1994), 'Experimental Analysis of Choice', *Marketing Letters* **5**, 351–368.

Carson, R.T., N.F. Meade and V.K. Smith (1993), 'Passive Use Values and Contingent Valuation: Introducing the Issues', *Choices* **8**, no. 2, 4–8.

Carson, R.T. and R.C. Mitchell (1993a), 'The Value of Clean Water: The Public's Willingness to Pay for Boatable, Fishable, and Swimmable Quality Water', *Water Resources Research* **29**, 2445–2454.

Carson, R.T. and R.C. Mitchell (1993b), 'The Issue of Scope in Contingent Valuation Surveys', *American Journal of Agricultural Economics* **75**, 1263–1267.

Carson, R.T. and R.C. Mitchell (1995a), 'Current Issues in the Design, Administration, and Analysis of Contingent Valuation Surveys', in P.O. Johansson, B. Kristrom, and K.G. Maler, eds., *Current Issues in Environmental Economics.* Manchester: Manchester University Press.

Carson, R.T. and R.C. Mitchell (1995b), 'Sequencing and Nesting in Contingent Valuation Surveys', *Journal of Environmental Economics and Management* **28**, 155–178.

Carson, R.T., R.C. Mitchell, W.M. Hanemann, R.J. Kopp, S. Presser, and P.A. Ruud, (1992), *A Contingent Valuation Study of Lost Passive Use Values Resulting from the Exxon Valdez* Oil Spill, Report to the Attorney General of the State of Alaska.

Carson, R.T., R.C. Mitchell, W.M. Hanemann, R.J. Kopp, S. Presser, and P.A. Ruud, (2003), 'Contingent Valuation and Lost Passive Use: Damages from the *Exxon Valdez* Oil Spill', *Environmental and Resource Economics* **25**, 257–286.

Carson, R.T. and D. Steinberg (1990), 'Experimental Design for Discrete Choice Voter Preference Surveys', in *1989 Proceeding of the Survey Methodology Section of the American Statistical Association.* Washington, DC: American Statistical Association.

Carson, R.T., L. Wilks and D. Imber (1994), 'Valuing the Preservation of Australia's Kakadu Conservation Zone', *Oxford Economic Papers* **46**, 727–749.

Champ, P.A. (1997), 'A Comparison of Contingent and Actual Voting Behavior', Paper presented at the annual U.S.D.A. W-133 Conference on Benefits and Costs in Natural Resource Planning, Portland, OR.

Champ, P.A., R.C. Bishop, T.C. Brown, and D.W. McCollum, (1997) 'Using Donation Mechanisms to Value Nonuse Benefits from Public Goods', *Journal of Environmental Economics and Management* 33, 151–162.

Chapman, R.G. and R. Staelin (1982), 'Exploiting Rank Ordered Choice Set Data Within the Stochastic Utility Model', *Journal of Marketing Research* **19**, 288–301.

Charles River Associates (1992), 'Methodological Biases in Valuing Environmental Resource Damage', *CRA Review*, December, 1–4.

Chilton, S.M and W.G. Hutchinson (1999), 'Some Further Implications of Incorporating the Warm Glow of Giving into Welfare Measures: A Comment on the Use of Donation Mechanism by Champ *et al.*', *Journal of Environmental Economics and Management* **37**, 202–209.

Ciriacy-Wantrup, S.V. (1947), 'Capital Returns from Soil-Conservation Practices', *Journal of Farm Economics* **29**, 1181–1196.

Collins, S. (1991), 'Prediction Techniques for Box-Cox Regression Models', *Journal of Business and Economics Statistics* **9**, 1991, 267–277.

Couch, A. and K. Keniston (1960), 'Yeasayers and Naysayers: Agreeing Response Set as a Personality Variable', *Journal of Abnormal and Social Psychology* **60** 151,–171.

Council of American Survey Research Organizations (1982), 'On the Definition of Response Rates', Report of the CASRO Task Force on Completion Rates. Port Jefferson, NY: Council of American Survey Research Organizations.

Davis, R.K. (1963), 'The Value of Big Game Hunting in a Private Forest', in *Transactions of the Twenty-Eighth North American Wildlife and Natural Resources Conference.* Washington, DC: Wildlife Management Institute.

Deacon, R. and P. Shapiro (1975), 'Private Preference for Collective goods Revealed Through Voting on Referenda," *American Economic Review* **65**, 943–955.

Deaton, A. and J. Muellbauer (1980), *Economics and Consumer Behavior*. New York: Cambridge University Press.

Desvousges, W.H., A.R. Gable, R.W. Dunford, and S.P. Hudson (1993), 'Contingent Valuation: The Wrong Tool for Damage Assessment', *Choices* **8**(2), 9–11.

Desvousges, W. H., F.R. Johnson, R.W. Dunford, K.J. Boyle, S.P. Hudson, and K.N. Wilson (1992), *Measuring Nonuse Damages Using Contingent Valuation: An Experimental Evaluation of Accuracy* (Research Triangle Park, NC: Research Triangle Institute).

Desvousges, W.H., M.C. Naughton, and G.R. Parsons (1992), 'Benefit Transfer – Conceptual Problems in Estimating Water Quality', *Water Resources Research* **28**, 675–683.

Diamond, P.A. and J.A. Hausman (1993), 'On Contingent Valuation Measurement of Nonuse Values', in J.A. Hausman, ed. *Contingent Valuation: A Critical Assessment*, Amsterdam: North-Holland.

Diamond, P.A. and J.A. Hausman (1984), 'Contingent Valuation: Is Some Number Better Than No Number?', *Journal of Economic Perspectives* **8**(4), 45–64.

Diamond, P.A., J.A. Hausman, G.K. Leonard, and M.A. Denning (1993), 'Does Contingent Valuation Measure Preferences', in J.A. Hausman, ed., *Contingent Valuation: A Critical Assessment*. Amsterdam: North-Holland.

Duffield, J. and D. Patterson (1992), 'Field Testing Existence Values: Comparison of Hypothetical and Cash Transactions Values', in B. Rettig, ed., Benefit and Costs in Natural Resource Planning. Fifth Interim Report. Corvallis, OR: Oregon State University, Department of Agriculture and Resource Economics.

Dunford, R.W., A.R. Gable, S.P. Hudson, W.H. Desvousges, and F.R. Johnson (1996), *A Review of the 1996 Contingent Valuation Study on the Value of Preventing Oil Spills Along California's Central Coast*. Durham, NC: Triangle Economic Research.

Dunford, R.W., K.E. Mathews and F.R. Johnson (1997), 'Conjoint Analysis: A Technique for Determining Compensatory Restoration', Paper presented at International Oil Spill Conference in Fort Lauderdale, Florida.

Efron, B. and R.J. Tibshirani (1993), *An Introduction to the Bootstrap*. London: Chapman and Hall.

Finney, D.J. (1949), 'The Estimation of the Parameters of Tolerance Distributions', *Biometrika* **36**, 239–256.

Flores, N.E. and R.T. Carson, (1997), 'The Relationship Between Income Elasticities of Demand and Willingness to Pay', *Journal of Environmental Economics and Management* **33**, 287–295.

Flores, N.E. and J. Thacher (2002), 'Money, Who Needs It? Natural Resource Damage Assessment', *Contemporary Economic Policy* **20**, 171–178.

Freeman, A.M., III (1993), *The Measurement of Environmental and Resource Values: Theory and Methods*. Washington, DC: Resources for the Future.

Greene, W.H., L. Knapp, and T.G. Seaks (1995), 'Estimating the Functional Form of the Independent Variables in Probit Models',. *Applied Economics* **27**, 193–196.

Grice, H. P. (1975), 'Logic and Conversation', in Cole, P. and J. Morgan, eds., *Syntax and Semantics 3: Speech Acts*. New York: Academic Press.

Haab, T.C. and K.E. McConnell (1997), 'Referendum Models and Negative Willingness to Pay: Alternative Solutions', *Journal of Environmental Economics and Management* **32**, 251–270.

Haab, T.C. and K.E. McConnell (1998), 'Referendum Models and Economic Values: Theoretical, Intuitive, and Practical Bounds on Willingness to Pay', *Land Economics* **74**, 216–229.

Haab, T.C. and K.E. McConnell (2002) *Valuing Environmental and Natural Resources: The Econometrics of Nonmarket Valuation*. Northhampton, MA: Edward Elgar.

Hanemann, W.M. (1984), 'Discrete/Continuous Models of Consumer Demand', *Econometrica* **52**, 541–561.

Hanemann, W.M. (1991), 'Willingness to Pay and Willingness to Accept: How Much Can They Differ?', *American Economic Review* **81**, 635–647.

Hanemann, W.M. (1992), 'Preface', in S. Navrud, ed., *Pricing the European Environment*. New York: Oxford University Press.

Hanemann, W.M. (1994), 'Valuing the Environment Through Contingent Valuation', *Journal of Economic Perspectives* **8**(4), 1994, 19–43.

Hanemann W.M. (1995), 'Contingent Valuation and Economics', in K.G. Willis and J.T. Corkindale, eds., *Environmental Valuation: New Perspectives*. Wallingford, Oxon, UK: CAB International, 79–117.

Hanemann, W.M. (1996), 'Theory Versus Data in the Contingent Valuation Debate', in D.J. Bjornstad and J.R. Kahn, eds., *The Contingent Valuation of Environmental Resources: Methodological Issues and Research Needs*. Brookfield, VT: Edward Elgar.

Hanemann, W.M. (1999), 'Neo-Classical Economic Theory and Contingent Valuation', in I.J. Bateman and K.G. Willis, eds., *Valuing Environmental Preferences: Theory and Practice of the Contingent Valuation Method in the US, EU, and Developing Countries*. New York: Oxford University Press.

Hanemann, W.M. and B.J. Kanninen (1999), 'The Statistical Analysis of Discrete-Response CV Data', in in I.J. Bateman and K.G. Willis, eds., *Valuing Environmental Preferences: Theory and Practice of the Contingent Valuation Method in the US, EC, and Developing Countries*. Oxford: Oxford University Press.

Harrison, G.W. (1992), 'Valuing Public Goods with the Contingent Valuation Method: A Critique of Kahneman and Knetsch', *Journal of Environmental Economics and Management* **23**, 248–257.

Harrison, G.W. and B. Kriström (1995), 'On the Interpretation of Responses in Contingent Valuation Surveys', in P.O. Johansson, B. Kriström and K.G. Mäler, eds., *Current Issues in Environmental Economics*. Manchester: Manchester University Press.

Hartigan, J. (1975), *Clustering Algorithms*. New York: Wiley.

Hausman, J.A., ed. (1993), *Contingent Valuation: A Critical Assessment*. Amsterdam: North-Holland.

Hausman, J.A. and W.K. Newey (1995), 'Nonparametric Estimation of Exact Consumers Surplus and Deadweight Loss', *Econometrica* **63**, 1445–1476.

Havemann, J. (1993), 'Why Oil Spills are Increasing', *Los Angeles Times*, Part A:1, 26 March.

Heckman, J. and B. Singer (1984), 'A Method for Minimizing the Impact of Distributional Assumptions in Econometric Models for Duration Data', *Econometrica* **52**, 271–320.

Imber, D., G. Stevenson and L. Wilks (1991), 'A Contingent Valuation Survey of the Kakadu Conservation Zone', Australian Resource Assessment Commission Research Paper No. 3. Canberra, Australia.

Johnson, F.R., and W.H. Desvousges (1997), 'Estimating Stated Preferences with Rated Pair Data: Environmental, Health and Employment Effects of Energy Programs', *Journal of Environmental Economics and Management* **34**, 79–99.

Johnson, F.R., D.J. MacNair, and E.E. Fries (1997), 'Ill-Considered Values: Contingent Valuation in Light of Evidence from Multiple-Response Stated-Preference Surveys', Triangle Economic Research, Working Paper No. T-9703.

Just, R. E., D. L. Hueth, and A.Schmitz (1982), *Applied Welfare Economics and Public Policy*, Englewood Cliffs, NJ: Prentice-Hall.

Kahneman, D. and J. Knetsch (1992), 'Valuing Public Goods: The Purchase of Moral Satisfaction', *Journal of Environmental Economics and Management* **22**, 57–70.

Kanninen, B.J. (1995), 'Bias in Discrete Response Contingent Valuation', *Journal of Environmental Economics and Management* **28**, 114–125.

Kaufman, L. and P.J. Rousseeuw (1990), *Finding Groups in Data: An Introduction to Cluster Analysis*. New York: Wiley.

Kealy, M.J., M. Montgomery and J.F. Dovidio (1990), 'Reliability and Predictive Validity of Contingent Values: Does the Nature of the Good Matter?,' *Journal of Environmental Economics and Management* **19**, 244–263.

Knetsch, J.L. and R.K. Davis (1966), 'Comparisons of Methods for Resource Evaluation', in A.V. Kneese and S.C. Smith, eds., *Water Research*. Baltimore: Johns Hopkins University Press.

Kristrom, B. (1990), 'A Non-Parametric Approach to the Estimation of Welfare Measures in Discrete Response Valuation Studies', *Land Economics* **66** 135–139.

Landes, W.M. and R.A. Posner (1987), *The Economic Structure of Tort Law*. Cambridge: Harvard University Press.

Lehmann, E.L. (1986), *Testing Statistical Hypotheses, 2nd ed.* New York: Wiley.

Lindsay, J.L., and L.M. Ryan (1998), 'Methods for Interval-Censored Data', *Statistics and Medicine* **17**, 219–238.

Lodge, M., K.M. McGraw and P. Stroh (1989), 'An Impression-Driven Model of Candidate Evaluation', *American Political Science Review* **83**, 397–419.

Lohmann, S. (1993), 'A Signaling Model of Informative and Manipulative Political Action', *American Political Science Review* **87**, 319–333.

Lohmann, S. (1994), 'Information Aggregation Through Costly Political Action', *American Economic Review*, 84, 518–530.

Lunn, T. (1986), 'Segmenting and Constructing Markets', in R.M. Worcester and J. Downham, eds., *Consumer Market Research Handbook*,. New York: North-Holland.

McConnell, K.E. (1990), 'Models for Referendum Data: The Structure of Discrete Choice Models for Contingent Valuation', *Journal of Environmental Economics and Management* **18**, 19–34.

McConnell, K.E. (1997), 'Does Altruism Undermine Existence Value?', *Journal of Environmental Economics and Management* **32**, 22–37.

Maddala, G.S. (1983), *Limited-Dependent and Qualitative Variables in Econometrics*. New York: Cambridge University Press.

Mathews, K., F.R. Johnson, R.W. Dunford and W.H. Desvousges (1995), 'The Potential Role of Conjoint Analysis in Natural Resource Damage Assessments', Working Paper, Triangle Economic Research.

Mercer Management Consulting, Inc. (1993), *Analysis of Oil Spill Costs and Financial Responsibility Requirements*. Report prepared for California Department of Fish and Game, Office of Oil Spill Prevention and Response. Sacramento.

Mitchell, R.C. and R.T. Carson (1989), *Using Surveys to Value Public Goods: The Contingent Valuation Method*. Baltimore: Johns Hopkins University Press.

Mitchell, R.C. and R.T. Carson (1995), 'Current Issues in the Design, Administration and Analysis of Contingent Valuation Surveys', in P.O. Johansson, B. Kriström and K.G. Mäler, eds., *Current Issues in Environmental Economics*. New York: Manchester University Press.

Morgan, B.J.T. (1992),. *Analysis of Quantal Response Data*. New York: Chapman and Hall.

Nelson, Wayne (1982), *Applied Life Data Analysis*. New York: Wiley.

Nisbett, R.E. and T.D. Wilson (1977), 'Telling More than We Know: Verbal Reports on Mental Processes', *Psychological Review* **84**, 1977, 231–259.

Portney, P.R. (1994), 'The Contingent Valuation Debate: Why Economists Should Care', *Journal of Economic Perspectives* **8**(4), 3–18.

Randall, A. (1994), 'A Difficulty with the Travel Cost Method', *Land Economics* **70**, 88–96.

Randall, A. (1993), 'Passive-Use Values and Contingent Valuation: Valid for Damage Assessment', *Choices* 8(2), 12–15.

Randall, A., B.C. Ives and C. Eastman (1974), 'Bidding Games for Valuation of Aesthetic Environmental Improvements', *Journal of Environmental Economics and Management* **1**, 132–149.

Shyamsundar, P. and R.A. Kramer (1996), 'Tropical Forest Protection: An Empirical Analysis of the Costs Borne by Local People', *Journal of Environmental Economics and Management* **31**, 129–144.

Schkade, D.A. and J.W. Payne (1994), 'How People Respond to Contingent Valuation Questions: A Verbal Protocol Analysis of Willingness to Pay for an Environmental Regulation', *Journal of Environmental Economics and Management* **26**, 88–109.

Seip, K. and J. Strand (1992), 'Willingness to Pay for Environmental Goods in Norway: A Contingent Valuation Study with Real Payment', *Environmental and Resource Economics* **2**, 91–106.

Silberman, J., D.A. Gerlowski and N.A. Williams (1992), 'Estimating Existence Value for Users and Nonsuers of New Jersey Beaches', *Land Economics* **68**, 225–236.

Smith, V.K. (1986), 'To Keep or Toss the Contingent Valuation Method', in R.G. Cummings, D.S. Brookshire and W.D. Schulze, eds., *Valuing Environmental Goods: An Assessment of the Contingent Valuation Method.* Totowa, NJ: Rowman and Allanheld.

Smith, V.K. (1992), 'Arbitrary Values, Good Causes, and Premature Verdicts', *Journal of Environmental Economics and Management* **22**, 71–89.

Stock, J.H. (1991), 'Nonparametric Policy Analysis: An Application to Estimating Hazardous Waste Clean-Air Benefits', in J. Powell and G. Tauchen, eds., *Nonparametric and Semi-parametric Methods in Economics and Statistics.* Cambridge: Cambridge University Press.

Stokes, L. and T. Belin (1998), 'What is a Margin of Error,' Pamphlet in the series *What is a Survey?*, Washington, DC: American Statistical Association.

Tetlock, P. (1983), 'Accountability and the Perseverance of First Impressions', *Social Psychology Quarterly* **45**, 74–83.

Tufte, Edward R. (1983), *The Visual Display of Quantitative Information.* Cheshire, CT: Graphics Press.

Turnbull, B.W. (1976), 'The Empirical Distribution Function with Arbitrarily Grouped, Censored and Truncated Data', *Journal of the Royal Statistical Society* **38**, 290–295.

United States Court of Appeals for the District of Columbia (1989), 'Ohio v. United States Department of Interior', 880 F.2nd 432 (D.C. Cir. 1989).

U.S. Environmental Protection Agency (1994), *President Clinton's Clean Water Initiative: Analysis of Benefits and Costs*, [EPA/800/R-94/002; NTIS PB94-154101]. Washington, DC: U.S. Environmental Protection Agency.

Varian, H. (1992), *Microeconomic Analysis*, 3rd. ed. New York: Norton.

Vossler, C.A., J. Kerkvliet, S. Polasky, and O. Gainutdinova (2003), 'Externally Validating Contingent Valuation: An Open-Space Survey and Referendum in Corvallis, Oregon,' *Journal of Economic Behavior and Organization* **51**, 261–277.

Westat, Inc. (1995), *State Policy Study: Main Study Interviewer's Manual.* Rockville, MD: Westat, Inc.

APPENDIX A
Main Study Survey Instrument

A.3 PDF Version

START INTERVIEW TIME: ☐☐:☐☐ A.M. P.M.

SECTION A

A-1. Let's start by talking for a moment about some issues in California. Some may not be important to you, others may be.

SHOW CARD A

First, (READ X'd ITEM). Is this issue not important at all to you personally, not too important, somewhat important, very important, or extremely important? (READ EACH ITEM, BEGINNING WITH X'd ITEM; CIRCLE ONE CODE FOR EACH; REPEAT ANSWER CATEGORIES AS NECESSARY.)

		NOT IMPORTANT AT ALL	NOT TOO IMPORTANT	SOMEWHAT IMPORTANT	VERY IMPORTANT	EXTREMELY IMPORTANT	N/S
( )	a. Improving education in California elementary and secondary schools	1	2	3	4	5	8
( )	b. Reducing air pollution in California cities	1	2	3	4	5	8
( )	c. Maintaining local library services	1	2	3	4	5	8
( )	d. Reducing crime	1	2	3	4	5	8
( )	e. Protecting coastal areas from oil spills	1	2	3	4	5	8
( )	f. Finding ways to reduce state taxes	1	2	3	4	5	8

A-2. The State of California spends tax money on many programs for many different purposes. I'm going to read a list of some of these programs. For each one, I would like you to tell me how important it is to <u>you</u> that the State continue to spend money on it.

SHOW CARD A AGAIN

First, (READ X'd ITEM). (READ EACH ITEM, BEGINNING WITH X'd ITEM; CIRCLE ONE CODE FOR EACH; REPEAT ANSWER CATEGORIES AS NECESSARY.)

		NOT IMPORTANT AT ALL	NOT TOO IMPORTANT	SOMEWHAT IMPORTANT	VERY IMPORTANT	EXTREMELY IMPORTANT	N/S
( )	a. Providing job-training for the unemployed	1	2	3	4	5	8
( )	b. Providing shelters for the homeless	1	2	3	4	5	8
( )	c. Protecting wildlife	1	2	3	4	5	8
( )	d. Providing lifeguards at state beaches	1	2	3	4	5	8
( )	e. Providing public transportation for Los Angeles	1	2	3	4	5	8
( )	f. Building new state prisons	1	2	3	4	5	8

These are just a few of the programs the State of California currently spends tax money on. Proposals are sometimes made to the State for new programs; but the State does not want to start any new programs unless taxpayers are willing to pay the additional cost for them.

One way for the State to find out about this is to give people like you information about a program so that you can make up your own mind about it. (STOP)

Your views are useful to State decision makers in deciding what, if anything, to do about a particular situation. (STOP)

In interviews of this kind, some people think that the program they are asked about is not needed; (STOP) others think that it is. We want to know what you think. (STOP)

A-3. Have you ever been interviewed before about whether the State should start a new program?

YES	1
NO	2
N/S	8

In the past, people have been asked about various types of programs. In this interview, I am going to ask you about a program that would prevent harm from oil spills off one part of the California coast. (STOP)

I will begin with important background information. Then I will ask you whether you think this particular program is worthwhile and why you feel the way you do. (STOP)

SHOW CARD B

Along the California coast, there are three different types of shoreline.

- The areas shown here in <u>green</u> are mostly saltwater marshes.
- The areas shown in <u>brown</u> are mostly rocky shoreline.
- And, the areas in <u>yellow</u> are mostly sandy beaches. (STOP)

A-4. Have you visited any of these three types of California shoreline in the last 12 months?

YES 1

NO .... 2 **(SKIP TO A-6)**
N/S .... 8 **(SKIP TO A-6)**

A-5. And, which ones are those? **(CIRCLE THOSE MENTIONED)**

SALTWATER MARSH 1
ROCKY SHORE 2
SANDY BEACHES 3

A-6. Each year, tankers and barges carrying oil make about 3,000 trips in and out of California harbors and along the Central Coast.

- <u>Large</u> oil tankers called super-tankers deliver their cargo to storage tanks and oil refineries in the San Francisco Bay and in the Greater Los Angeles area.
- <u>Small</u> tankers and barges transport various types of refined oil back and forth along the 500 miles of coastline between San Francisco and the L.A. area.

SHOW CARD B AGAIN

Tankers and barges occasionally run into things like underwater rocks, other ships, or pipelines, and spill some of their oil into the water.

> Unless the spill is very small, the oil can harm wildlife.
>
> After an oil spill, the company that caused it must pay to clean up as much oil as possible from both the water and the shoreline. STOP

Over the years, the State has taken various steps to prevent harm from oil spills. Recently, steps have been taken to set up programs to prevent harm from spills in the San Francisco Bay and in the L.A. area. The State wants to know whether people think this would be worth doing for the Central Coast. STOP

> As you can see here, most of the Central Coast is rocky shoreline with some scattered sandy beaches.

Oil spills that harmed wildlife have happened here every few years. State and university scientists were asked to provide information about the effects of these past spills.

SHOW CARD C

This drawing shows the types of wildlife that Central Coast spills have harmed. It shows five types of birds and other types of small animals that live in or near the water. Take your time to look it over.

STOP UNTIL R IS FINISHED LOOKING AT CARD

SHOW CARD C AGAIN

The five birds shown here are the types of birds that past spills have harmed the most.

A-7. Do you happen to be familiar with any of these birds?

YES 1

NO .... 2 **(SKIP TO A-9)**
N/S .... 8 **(SKIP TO A-9)**

A-8. Which ones? **(CIRCLE THOSE MENTIONED)**

WESTERN GULL 1
PACIFIC LOON 2
RHINOCEROS AUKLET 3
COMMON MURRE 4
BRANDT'S CORMORANT 5

A-9. According to scientists, none of these birds are in any danger of becoming extinct.

The number next to each bird shows how many of them live in California. For example, there are about 290,000 Pacific Loons and 130,000 Western Gulls.

All five types of birds also live in other States. (STOP)

BOX 1

IF R ASKS IF WESTERN GULLS ARE THE SAME AS SEA GULLS, CHECK HERE ☐ AND SAY:

Western gulls are one of a dozen types of sea gulls.

SHOW CARD C AGAIN

Whenever oil washes up on the shoreline along the Central Coast, it harms many small animals and saltwater plants. Some are shown here.

They include clams, sea stars, crabs, mussels, kelp, and other seaweed.

None of these are in any danger of becoming extinct. (STOP)

Marine mammals—such as whales, seals, and dolphins—are not usually affected by the oil because they generally leave the area when a spill occurs. Fish also leave the area and are not affected. (STOP)

BOX 2

IF R ASKS WHAT HAPPENS TO SEA OTTERS OR MENTIONS HE/SHE THOUGHT SEA OTTERS WERE ALSO AFFECTED IN OIL SPILLS, CHECK HERE ☐ AND SAY:

Like other marine mammals, sea otters usually leave the area when a spill occurs. They have not usually been affected by past Central Coast spills.

Recently, the federal government passed a new law to help reduce the number of oil spills.

> Ten years from now, all oil tankers and barges will be required to have two outer hulls instead of the single-hull most of them have now. Double-hulls provide much more protection against oil leaking after an accident.

However, it will take ten years before all single-hulled tankers and barges can be replaced. Until then, spills are expected to happen every few years along the Central Coast, just as they have in the past, unless something is done.

SHOW CARD D

This shows the total amount of harm to wildlife that state and university scientists expect will happen in the Central Coast area over the next ten years. It is based on studies scientists have made of past spills in this area.

SHOW CARD D AGAIN

In the next ten years:

- scientists expect that a total of about 12,000 birds of various types will be killed by oil spills off the Central Coast.

  In addition, about 1,000 more birds are expected to be injured but survive.

- Also, many small animals and saltwater plants are likely to be killed along a total of about ten miles of shoreline. (STOP)

The harm from an oil spill is not permanent. Over time, waves and other natural processes break down the oil in the water and on the shoreline.

- Typically, within ten years or less after a spill, there will be as many of the affected birds as before the spill.

  The small animals and saltwater plants in the affected area recover somewhat faster, in about five years or less. (STOP)

A-10. Is there anything more that you would like to know about the harm oil spills are expected to cause off the Central Coast over the next ten years?

YES 1

NO .... 2 **(SKIP TO A-11)**

A-10A. What is that?

____

____

____

____

A-11. If taxpayers think it is worthwhile, the State could prevent this harm by setting up a prevention program for this part of the coast. This program would be similar to those successfully used by other states, such as the State of Washington. It would last for ten years, until all tankers and barges have double-hulls.

This program would do two things.

First, it would help prevent oil spills from occurring.

Second, if an oil spill does occur, it would prevent the oil from spreading and causing harm.

Here is how a Central Coast program would prevent spills from occurring.

SHOW CARD E

- Oil spill prevention and response centers would be set up in three different locations along this part of the coast.

  Specially-designed ships, called escort ships, would be based at each center.

- An escort ship would travel alongside every tanker and barge as it sails along the Central Coast. This would help prevent spills in this area by keeping the tankers and barges from straying off-course and running into underwater rocks, other ships, or pipelines.

SHOW CARD F

If any oil were spilled, here's how the program would keep it from spreading and causing harm.

- The crew of the escort ship would quickly put a large floating sea fence into the water to surround the oil. To keep it from spreading in rough seas, this fence would extend 6 feet above and 8 feet below the surface of the water.

- Then skimmers, like the one shown here, would suck the oil from the surface of the water into storage tanks on the escort ship.

  Other ships would be sent from the nearest prevention and response center to aid in the oil recovery and clean-up. (STOP)

A-12. Is there anything more you would like to know about how this prevention program would work?

YES 1

NO .... 2 **(SKIP TO A-13)**

A-12A. What is that?

____

____

____

BOX 3

IF R ASKS ABOUT HOW PROGRAM WOULD BE PAID FOR OR ABOUT PROGRAM COST, CHECK HERE ☐ AND SAY:

I will come to that in just a moment.

A-13. The money to pay for this program would come from both the tax-payers and the oil companies. Because individual oil companies cannot legally be required to pay the cost of setting up the program, all California households would pay a special one time tax for this purpose.

> This tax money would pay for providing the escort ships and setting up the three oil spill prevention and response centers along the Central Coast.

Once the prevention program is set up, all the expenses of running the program for the next ten years would be paid by the oil companies.

> This money would come from a special fee the oil companies would be required to pay each time their tankers and barges were escorted along the Central Coast.

Once the federal law goes into effect ten years from now, all tankers and barges will have double-hulls and this program would be closed down. (STOP)

BOX 4

IF R ASKS ABOUT PROGRAM COST, CHECK HERE ☐ AND SAY:

> I will come to that in just a moment.

IF R SAYS OIL COMPANIES SHOULD PAY ALL COSTS, CHECK HERE ☐ AND SAY:

> The State cannot legally force individual oil companies to pay for setting up the program. However, the oil companies can be required to pay a special fee each time one of their ships is escorted along the Central Coast. These fees will pay to keep the program operating over the next ten years.

We are interviewing people to ask how they would vote on this Central Coast prevention program if it were put on the ballot in a California election. (STOP)

There are reasons why you might vote for setting up this program and reasons why you might vote against it.

SHOW CARD G

The program would prevent harm from oil spills in the Central Coast area during the next ten years. Specifically, the program would:

- prevent the deaths of about 12,000 birds as well as the deaths of many small animals and saltwater plants along about 10 miles of shoreline, and

  prevent 1,000 more birds from being injured. (STOP)

On the other hand,

- the number of birds and other wildlife it would protect is small in comparison to their total numbers, and none are endangered. (STOP)

  Your household might prefer to spend the money to solve other social or environmental problems instead. (STOP)

  Or, the program might cost more than your household wants to spend for this.

REMOVE CARD G

SECTION B

If the Central Coast prevention program were put into place, it would cost your household a total of $5. You would pay this as a special one time tax added to your next year's California income tax.

B-1. If an election were being held today, and the total cost to your household for this program would be $5, would you vote for the program or would you vote against it?

FOR 1

AGAINST .... 2 **(SKIP TO B-4)**

N/S .... 8 **(SKIP TO B-5)**

BOX 5

IF R SAYS OIL COMPANIES SHOULD PAY ALL COSTS, CHECK HERE ☐ AND SAY:

The State cannot legally force individual oil companies to pay for setting up the program. However, the oil companies can be required to pay a special fee each time one of their ships is escorted along the Central Coast. These fees will pay to keep the program operating over the next ten years.

B-2. People have different reasons for voting for the Central Coast prevention program. What would the program do that made you willing to pay for it? **(PROBE: Was there something specific that the program would do that made you willing to pay for it?)**

____

____

____

____

____

B-3. Occasionally, people vote for the program because they are concerned that oil spills may somehow harm human health. Suppose human health was definitely not affected and the program would only prevent harm to birds, small animals, and saltwater plants. Would you vote for or against the program if it cost your household $5?

VOTE "FOR" 1

VOTE "AGAINST" 2

N/S 8

SKIP TO SECTION C

B-4. Did you vote against the program because it isn't worth that much money to you, or because it would be somewhat difficult for your household to pay that amount, or because of some other reason?

ISN'T WORTH THAT AMOUNT 1

DIFFICULT TO PAY 2

OTHER REASON (SPECIFY) 3

____

____

____

SKIP TO SECTION C

B-5. Could you tell me why you aren't sure about how you would vote? (**PROBE**)

____

____

____

____

SECTION C

Please think back to a few moments ago when I asked you whether you would vote for or against the program.

SHOW CARD H

C-1. At that time, did you think the harm from oil spills in the Central Coast over the next ten years would be about the <u>same</u> as that shown here, or a lot <u>more</u> or a lot <u>less</u>?

SAME 1
A LOT MORE 2
A LOT LESS 3
OTHER (**SPECIFY**) 4

____

____

____

NS 8

SHOW CARD I

C-2. How serious did you consider this amount of harm to be . . .

<u>Not</u> serious at all 1
Not <u>too</u> serious 2
<u>Somewhat</u> serious 3
<u>Very</u> serious, or.. 4
<u>Extremely</u> serious? 5
N/S 8

SHOW CARD J

C-3. Did it seem to you that the prevention program I told you about would be completely effective at preventing harm from Central Coast oil spills, mostly effective, somewhat effective, not too effective, or not effective at all?

COMPLETELY EFFECTIVE............................1
MOSTLY EFFECTIVE....................................2
SOMEWHAT EFFECTIVE 3
NOT TOO EFFECTIVE....................................4
NOT EFFECTIVE AT ALL..............................5
N/S 8

C-4. When you decided how to vote, did you think your household would have to pay the special tax for the program for one year or for more than one year?

ONE YEAR..1
MORE THAN ONE YEAR 2
N/S 8

C-5. Thinking about everything I have told you during this interview, overall did it try to push you to vote one way or another, or did it let you make up your own mind about which way to vote?

PUSHED ONE WAY OR ANOTHER........................1

LET ME MAKE UP OWN MIND ....2 **(SKIP TO SECTION D)**
N/S...8 **(SKIP TO SECTION D)**

C-6. Which way did you think it pushed you?

VOTE FOR THE PROGRAM....................................1
VOTE AGAINST THE PROGRAM...........................2
OTHER **(SPECIFY)**...3

____

____

____

N/S...8

C-6A. What was it that made you think that? **(PROBE: "Can you be more specific about what you have in mind?" "Anything else?")**

____

____

____

SECTION D

Now I would like to ask you a few questions about your household's recreational activities.

D-1. Has anyone in your household ever driven along the Central Coast on Highway 1, the coast highway?

YES 1

NO .... 2 **(SKIP TO D-3)**
N/S .... 8 **(SKIP TO D-3)**

D-2. And, was this in the last five years?

YES 1
NO 2
N/S 8

D-3. In the past five years, has anyone in your household gone saltwater boating or saltwater fishing?

YES 1
NO 2
N/S 8

D-4. Does anyone in your household like to identify different species of birds?

YES 1
NO 2
N/S 8

D-5. During this past summer, about how many times did people in your household go to beaches anywhere along the California coast . . .

Never, ... 1
Once or twice, ... 2
Three to ten times, or.......... 3
More than ten times? 4
N/S 8

SHOW CARD K

D-6. How often do you personally watch television programs about animals and birds in the wild . . .

Very often, ... 1
Often, ... 2
Sometimes, ... 3
Rarely, or 4
Never? 5
N/S 8

SHOW CARD L

D-7. Do you think of yourself as an . . .

Environmental activist, a 1
Strong environmentalist, a 2
Somewhat strong environmentalist, a 3
Not particularly strong environmentalist, or 4
Not an environmentalist at all? 5
N/S 8

Now, just a few questions about your background.

D-8. First, in total, how many years have you lived in California?

☐☐ YEARS

D-9. In what month and year were you born?

☐☐/☐☐
MONTH YEAR

D-10. What is the <u>highest</u> year of school you completed or the highest degree you received?

THROUGH 8th GRADE 01
9th, 10th, 11th, 12th GRADE (NO DIPLOMA) 02
HIGH SCHOOL EQUIVALENT (for example, GED) 03
HIGH SCHOOL GRADUATE (DIPLOMA) 04
SOME COLLEGE BUT NO DEGREE 05
ASSOCIATES DEGREE IN OCCUPATIONAL
OR VOCATIONAL PROGRAM 06
ASSOCIATES DEGREE IN ACADEMIC
PROGRAM 07
BACHELOR'S DEGREE (for example, BA, AB, BS) 08
MASTER'S DEGREE (for example, MA, MS, MEng,
MEd, MSW, MBA) 09
PROFESSIONAL SCHOOL DEGREE (for example, MD,
DDS, DVM, LLB, JD) 10
DOCTORATE DEGREE (for example, PhD, EdD) 11
REFUSED 97
N/S 98

D-11. Currently, how many adults in your household, <u>including yourself</u>, work for pay?

0 1 2 3 4 5 6 (OR MORE)

NUMBER WHO CURRENTLY WORK FOR PAY

SHOW CARD M

D-12. I'd like you to think about the income received last year by everyone in your household. Adding together all income for everyone in your household, which letter on this card best describes your household's total income for last year—1994—before taxes? Please include wages or salaries, social security or other retirement income, child support, public assistance, business income, and all other income.

LETTER A	01	
LETTER B	02	
LETTER C	03	**(SKIP TO D-14)**
LETTER D	04	**(SKIP TO D-14)**
LETTER E	05	**(SKIP TO D-14)**
LETTER F	06	**(SKIP TO D-14)**
LETTER G	07	**(SKIP TO D-14)**
LETTER H	08	**(SKIP TO D-14)**
LETTER I	09	**(SKIP TO D-14)**
LETTER J	10	**(SKIP TO D-14)**
LETTER K	11	**(SKIP TO D-14)**
N/S	98	**(SKIP TO D-14)**
REFUSED	97	**(SKIP TO D-14)**

D-13. Did anyone in your household pay any California income taxes for last year, 1994, by having taxes withheld from wages, retirement income, or other money received, or has anyone in your household sent, or intend to send, tax money for last year to the State with a tax form?

YES	1
NO	2
N/S	8

SHOW CARD N

D-14. When you look ahead to the next few years, do you see your personal financial situation getting . . .

Much better, ... 1
A little better, ... 2
Staying about the same, ... 3
Getting a little worse, or 4
Much worse? 5
OTHER 6
N/S 8

D-15. Now that we're almost at the end of the interview and you have been able to think a bit more about the situation, I'd like to give you a chance to review your answer to the voting question.

You were asked if you would vote for or against a program that would prevent the harm that I showed you earlier on this card.

SHOW CARD O

If an election were being held today, would you vote for the program or against the program if it cost your household a one-time tax payment of $5?

FOR 1
AGAINST 2
N/S 8

D-16. There are different ways for people to pay for new programs to protect the environment.

One way is for the government to pay the cost. This will raise everyone's <u>taxes</u>.

Another way is for businesses to pay the cost. This will make <u>prices</u> go up for everyone.

If you <u>had</u> to choose, would you prefer to pay for new environmental programs ...

Through higher taxes, or 1
Through higher prices? 2
EITHER ONE/DON'T CARE WHICH 3
NEITHER 4
N/S 8

SHOW CARD P

D-17. Generally speaking, how much confidence do you have in the California state government? Would you say . . .

A great deal, ... 1
Some, ... 2
Hardly any, or 3
None? 4
N/S 8

END INTERVIEW TIME: ☐☐:☐☐ A.M. P.M.

D-18. What is your full name and phone number, in case my supervisor wants to check my work? (RECORD FULL NAME AND PHONE NUMBER ON RECORD OF ACTIONS. **DO NOT RECORD IT HERE**.)

RECORDED ON RECORD OF ACTION 1
NO PHONE 2
REFUSED 7

BOX S

IS THE SCREENER FOR THIS CASE COMPLETED?

YES	1	THANK RESPONDENT FOR COOPERATION.
NO	2	SAY TO RESPONDENT: "I have just a few more questions I need to ask about the other adults in your household. Let me verify that there are (number from AS-1) people 18 or older living in this household." RETURN TO SCREENER (PG 2). ASK QUESTIONS S-3 THROUGH S-5, ENUMERATION TABLE, AND S-12. THEN, THANK RESPONDENT FOR COOPERATION.

SECTION E
INTERVIEW EVALUATION QUESTIONS

PLEASE NOTE THE FOLLOWING ABOUT THE RESPONDENT BY CIRCLING THE NUMBER OF THE CORRECT RESPONSE:

E-1.	SEX	MALE	1
		FEMALE	2
E-2.	RACE	WHITE, NOT HISPANIC	1
		WHITE, HISPANIC	2
		BLACK, NOT HISPANIC	3
		BLACK, HISPANIC	4
		ASIAN	5
		OTHER (SPECIFY)	6
		______________________________	

E-3. TRANSFER THE RESPONDENT'S ZIP CODE FROM THE ADDRESS LABEL ON THE CALL RECORD FOLDER: ☐☐☐☐☐

E-4. What was the reaction of the respondent as you read **A-3 through A-13?** (This is the descriptive material including the maps and drawings).

		EXTREMELY	VERY	SOMEWHAT	SLIGHTLY	NOT AT ALL	N/S
a.	How distracted was the respondent?	1	2	3	4	5	8
b.	How attentive was the respondent?	1	2	3	4	5	8
c.	How interested was the respondent?	1	2	3	4	5	8

E-5. Did the respondent say anything suggesting that he or she had any difficulty understanding either the harm caused by Central Coast oil spills or the prevention program?

YES 1

NO.. 2 **(SKIP TO E-6)**

E-5A. Please describe the difficulties.

____

____

____

E-6. Did the respondent have any difficulty understanding the voting question, B-1?

YES...1

NO ....2 **(SKIP TO E-7)**

E-6A. Please describe the difficulties.

____

____

____

E-7. When you asked B-1, did you feel the respondent was impatient to finish the interview?

YES...1

NO ....2 **(SKIP TO E-8)**

NOT SURE...8 **(SKIP TO E-8)**

E-7A. How impatient was the respondent?

VERY IMPATIENT...............................1
SOMEWHAT IMPATIENT...................2
A LITTLE IMPATIENT 3
NOT VERY IMPATIENT......................4
NOT SURE...8

E-8. How serious was the consideration the respondent gave to the decision about how to vote?

EXTREMELY SERIOUS 1
VERY SERIOUS...................................2
SOMEWHAT SERIOUS........................3
SLIGHTLY SERIOUS...........................4
NOT AT ALL SERIOUS 5
NOT SURE...8

E-9. Not counting you and the respondent, was <u>anyone</u> age 13 or older present when the respondent voted?

YES 1

NO .... 2 **(SKIP TO E-10)**

OTHERS CAME IN AND OUT 3

E-9A. Do you think the other person(s) affected how the respondent voted or don't you know?

YES 1
NO 2
DON'T KNOW 8

E-10. Do you have any other comments about this interview?

____

____

____

____

APPENDIX A
Main Study Survey Instrument

A.4 Visual Aids

1. Not important at all
2. Not too important
3. Somewhat important
4. Very important
5. Extremely important

CARD A

In the visual aids binder used in the field, the facing page was blank.

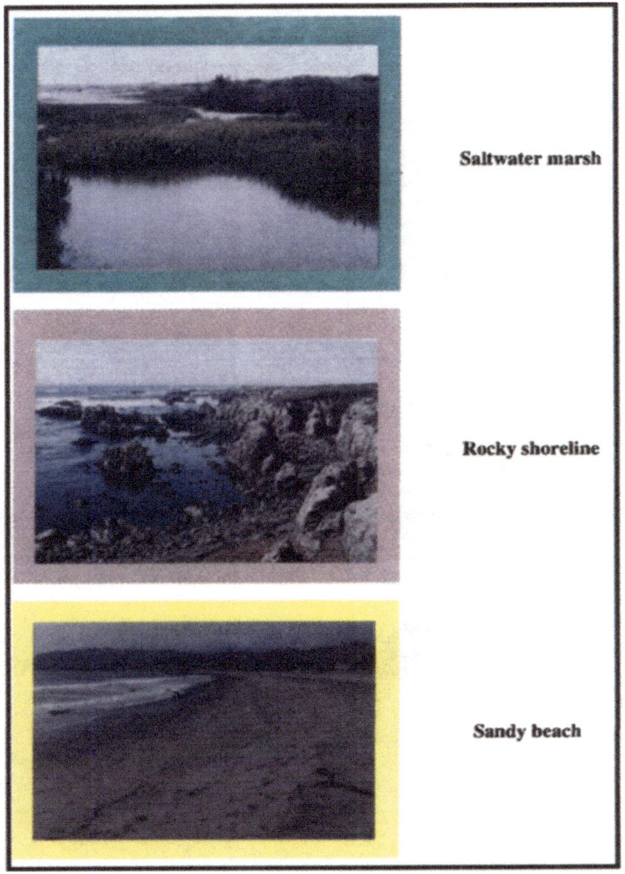

TOP OF CARD B

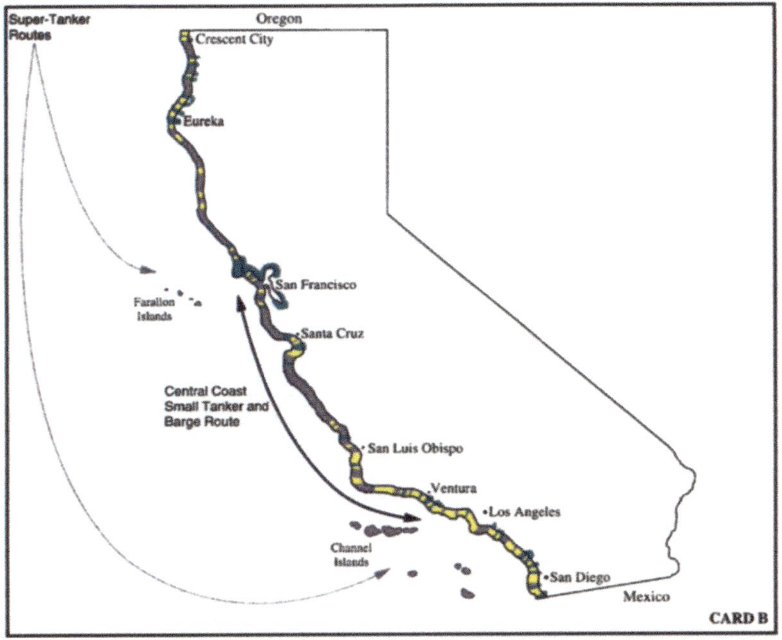

BOTTOM OF CARD B

In the visual aids binder used in the field, the two parts of Card B appeared on facing pages

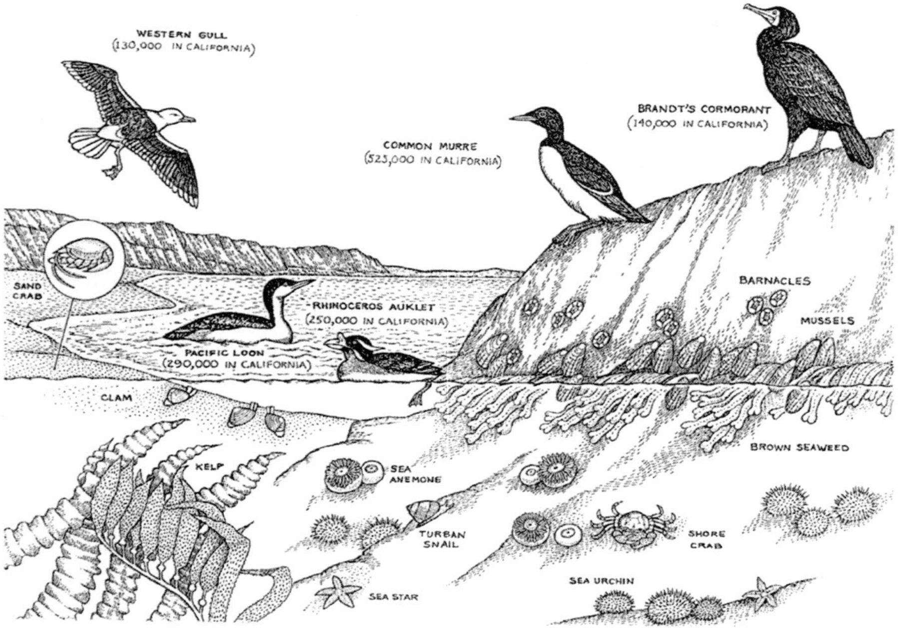
WESTERN GULL
(130,000 IN CALIFORNIA)
COMMON MURRE
(525,000 IN CALIFORNIA)
BRANDT'S CORMORANT
(140,000 IN CALIFORNIA)
SAND CRAB
RHINOCEROS AUKLET
(250,000 IN CALIFORNIA)
PACIFIC LOON
(290,000 IN CALIFORNIA)
BARNACLES
MUSSELS
CLAM
BROWN SEAWEED
KELP
SEA ANEMONE
TURBAN SNAIL
SHORE CRAB
SEA URCHIN
SEA STAR

CARD C

TOP OF CARD D

Expected Harm from Oil Spills off Central Coast

TOTAL HARM OVER NEXT 10 YEARS	TIME TO RECOVER AFTER EACH SPILL
12,000 birds killed **1,000 birds** injured	10 years
Many small animals and saltwater plants killed along **10 miles** of shoreline	5 years

CARD D

BOTTOM OF CARD D

In the visual aids binder used in the field, the two parts of Card D appeared on facing pages.

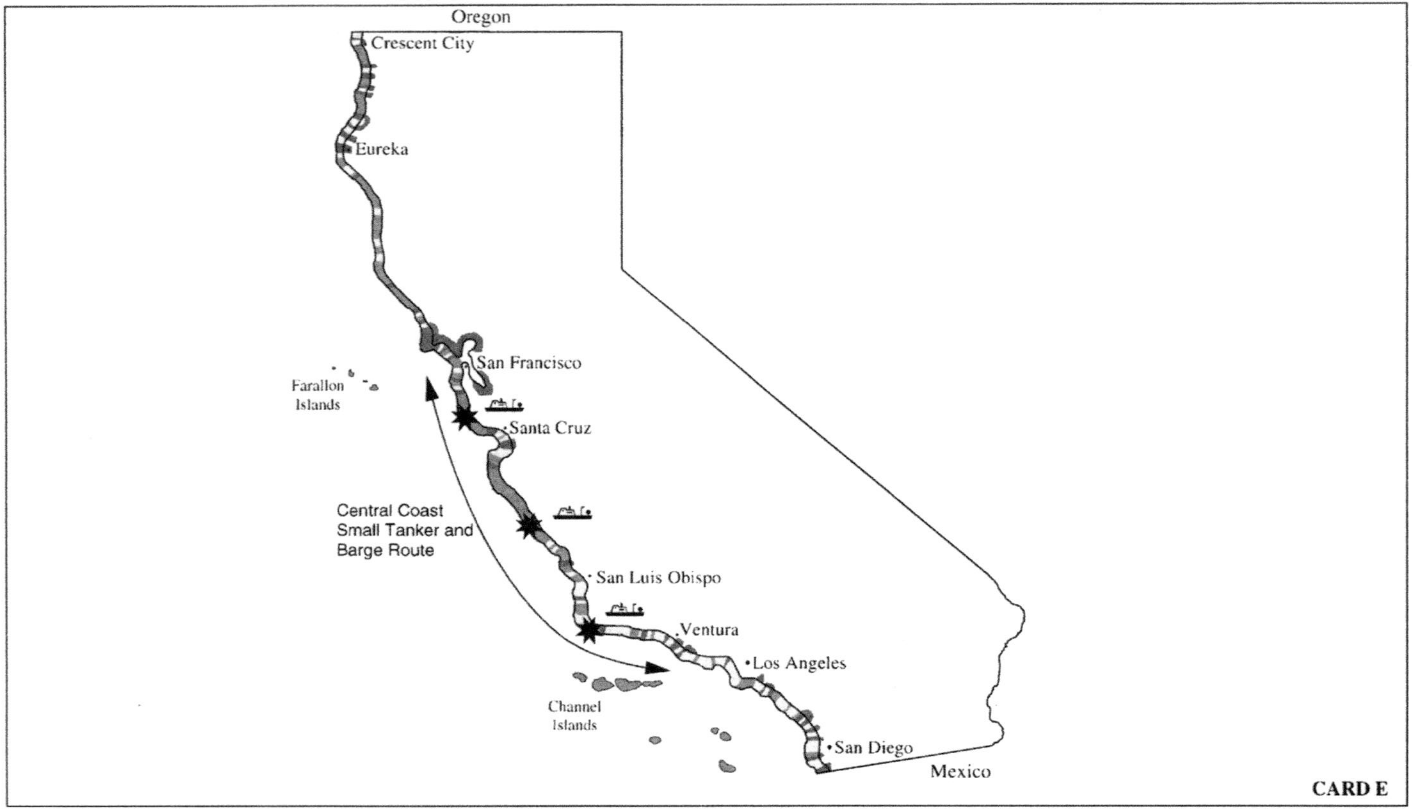

In the visual aids binder used in the field, this graphic was colored identically to that of CARD B, and the facing page was blank.

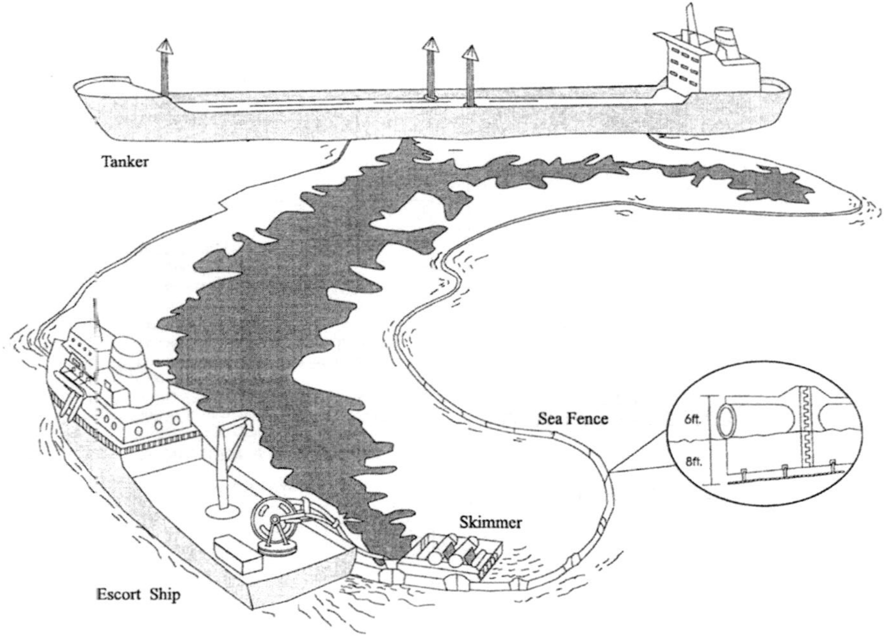

CARD F

In the visual aids binder used in the field, the facing page was blank.

REASONS *FOR*

Over the next 10 years, the Central Coast prevention program would:

prevent deaths of about **12,000 BIRDS** as well as deaths of many small animals and saltwater plants along about **10 MILES** of shoreline, and

prevent **1,000 MORE BIRDS** from being injured.

REASONS *AGAINST*

The **NUMBER** of birds and other wildlife the program would protect is **SMALL** in comparison to their **TOTAL** numbers, and none are endangered.

Your household might prefer to spend the money on **OTHER PROBLEMS** instead.

The program might **COST MORE** than your household wants to spend for this.

CARD G

Expected Harm from Oil Spills off Central Coast

TOTAL HARM OVER NEXT 10 YEARS	TIME TO RECOVER AFTER EACH SPILL
12,000 birds killed **1,000 birds** injured	10 years
Many small animals and saltwater plants killed along **10 miles** of shoreline	5 years

CARD H

In the visual aids binder used in the field, each card was alone on the page and the facing page was blank.

Expected Harm from Oil Spills off Central Coast

TOTAL HARM OVER NEXT 10 YEARS	TIME TO RECOVER AFTER EACH SPILL
12,000 birds killed **1,000 birds** injured	10 years
Many small animals and saltwater plants killed along **10 miles** of shoreline	5 years

TOP OF CARD I

1. Not serious at all
2. Not too serious
3. Somewhat serious
4. Very serious
5. Extremely serious

CARD I

BOTTOM OF CARD I

In the visual aids binder used in the field, the two parts of CARD I appeared on facing pages.

1. Completely effective
2. Mostly effective
3. Somewhat effective
4. Not too effective
5. Not effective at all

CARD J

1. Very often
2. Often
3. Sometimes
4. Rarely
5. Never

CARD K

In the visual aids binder used in the field, each card was alone on the page and facing page was blank.

1. Environmental activist

2. Strong environmentalist

3. Somewhat strong environmentalist

4. Not particularly strong environmentalist

5. Not an environmentalist at all

CARD L

Total Before-Tax Income in 1994 for Everyone in Your Household

(Please include all income sources, including social security or other retirement, etc., for everyone in your household.)

A. Under $10,000
B. $10,000 to $19,999
C. $20,000 to $29,999
D. $30,000 to $39,999
E. $40,000 to $49,999
F. $50,000 to $59,999
G. $60,000 to $69,999
H. $70,000 to $79,999
I. $80,000 to $99,999
J. $100,000 to $149,999
K. $150,000 or more

CARD M

In the visual aids binder used in the field, each card was alone on the page and facing page was blank.

1. Much better

2. A little better

3. About the same

4. A little worse

5. Much worse

CARD N

Expected Harm from Oil Spills off Central Coast

TOTAL HARM OVER NEXT 10 YEARS	TIME TO RECOVER AFTER EACH SPILL
12,000 birds killed **1,000 birds** injured	10 years
Many small animals and saltwater plants killed along **10 miles** of shoreline	5 years

CARD O

In the visual aids binder used in the field, each card was alone one the page and the facing page was blank.

1. Great deal

2. Some

3. Hardly any

4. None

CARD P

In the visual aids binder used in the field, the facing page was blank.

APPENDIX D
Coding Categories for Open-Ended Questions

A-10A What [more would you like to know about the harm oil spills are expected to cause off the Central Coast over the next ten years]?

1. **Some aspect of HARM NOT DESCRIBED.**

   "How do fish and others know how to leave the area?" "How are the birds killed?" "How found down in the sediment does the oil go?" "How long does it take to break oil down"?

2. **Possible impacts on HUMANS/HUMAN HEALTH/ DRINKING WATER.**

   "What about effects on humans?" "Does it effect the food people eat?" "What's going to happen to water supplies?"

3. **Possible impacts on ECONOMY/RECREATIONAL USE/ TOURISM.**

   "How does it affect the economy?" "How about the beaches themselves, people can't use them?" "What's the tourist impact?"

4. **VALIDITY/SOURCE of information.**

   "What have the scientists based their studies on?" "Who conducted the study for this information (about expected harm)?" "How accurate are these studies?"

5. **SPONSOR of survey.**

   "Is one group doing this survey?" "Are the oil companies paying for any cost of this survey?"

6. **COST of program; PAYING for the program.**

   "Who would pay?" "What is the cost?" "How much would we have to pay?"

7. **What can be done to prevent harm?**

   "What can be done in the meantime?" "How to help them (affected wildlife)?" "Is there an alternate way to transport oil?" "Besides the double hull, can't they partition other parts?"

8. **Some aspect of HARM ALREADY DESCRIBED.**

   "How long for the effect on birds?" So, the saltwater plants recover in 10 years?" "What effect does it have on fishes?"

9. **COMMENT/not a question.**

   "I understand this, but more important is crime." "I believe the study here is done by the oil companies."

10. **Other questions.**

    "Are they going to increase shipping in the next 10 years?"

A-12A What [more would you like to know about how this prevention program would work]?

1. **COST of program; PAYING for the program.**

   "Who would pay?" "What is the cost?" "How much would we have to pay?"

2. **ADDITIONAL information about specific PROGRAM FEATURE.**

   "How fast is the response of the ships?" "How long does (the sea fence) take to set up?" "Who would maintain the (escort) ships?"

3. **Information about program feature ALREADY DESCRIBED.**

   "Every time a tanker goes out, it would be escorted?" "Is there more than one escort ship per boat?" "When will (the program) start?" "So, this program has a finite end?"

4. **OTHER WAYS to prevent harm/ALTERNATIVES.**

   "Couldn't they make the tankers go out further from the coast to avoid collisions?" "Why don't they make smaller ships with less oil?" "Why can't the tanker carry this equipment?"

5. **What happens to SPILLED/RECOVERED OIL?**

   "Does the oil stay on top?" "What happens to the oil once it's on the escort ship?" "What are they going to do with the oil?"

6. **Whether program would WORK/past experience with.**

   "Would it work?" "Is it effective?" "How effective is it if the weather is stormy?" "How successful was this program in the past?"

7. **COMMENT/not a question.**

   "I think it's a good idea." "Sounds like you have explained it well."

8. **Other questions.**

   "Is the excess of other ships in the water going to cause more problems?" "Once a leak begins in a tanker, can you stop it?"

B-2 People have different reasons for voting for the Central Coast prevention program. What would the program <u>do</u> that made <u>you</u> willing to pay for it?

1. **Program would PROTECT THE WILDLIFE and/or the potentially AFFECTED ENVIRONMENT**

   Program will help some or all of the animals affected by the spills. "To help/save the wildlife/animals." "To save the wildlife/animals you told me about." "Prevent death of animals/birds." "It saves the wildlife." "Keeps more animals living."

   Protect the shoreline/ocean. "Stop harm to shoreline." "It would keep the water clean." "Help clean-up oil spill." "Keep beaches/shoreline clean."

2. **Program would help <u>OTHER SPECIFIC</u> ANIMALS besides those described.**

   "I don't feel all the fish swim away." "Would prevent harm to sea otters and other marine mammals."

3. **Program would prevent PERMANENT damage.**

   "Would prevent permanent scar on landscape." "There could be extinction in the future."

4. **Program would protect the ENVIRONMENT (IN GENERAL).**

   "For environmental purposes." "It's saving the ecosystem."

5. **Program would prevent PHYSICAL HARM to PEOPLE (including respondent).**

   "It would protect our drinking water." "Prevent health hazard to swimmers."

6. **RESPONDENT (including his or her household) is personally CONCERNED about the environment or would BENEFIT from the program personally in ways <u>other than</u> health.**

   "*I'm* an environmentalist/animal activist/concerned about wildlife." "*I* like animals." "*I* fish/am a sportsman/an outdoors person." "*I* would feel better knowing there wouldn't be any more oil spills." [Italics added.]

7. **Program will benefit OTHER PEOPLE (i.e., not the respondent) in ways other than health.**

   Program would make it possible for others to enjoy animals. "Make it possible for people to see/enjoy the animals."

   Program would benefit future generations. "Help future generations enjoy the ocean."

   Program would help people use the local environment. "Make it possible for people to see/enjoy/use the ocean/shoreline."

8. **PEOPLE (collectively) CAUSED the problem and/or are RESPONSIBLE for fixing it.**

   "Make up for *our* past mistakes." "*We* need to protect/clean-up the environment/the world/nature/ecosystem." "Evidence that we're taking responsibility for big businesses' damage." [Italics added.]

9. **The COST of the program is affordable/reasonable.**

   "Cost is cheap." "Small price to pay for the environment." "Fixing it up later will cost more."

10. **Program would WORK** (*i.e.*, something intrinsic to the program itself).

    "I thought it was a good plan and thought it would work." "Because the program is proactive and preventative." "Escort ships would help stop the pollution." "Response time would be shorter."

11. **Program would make the OIL COMPANIES more responsible.**

    "It would make the oil companies function responsibly."

12. **OTHER REASONS.**

    "I don't know." Not sure if it will really work but willing to give it a try" etc.

B-4 Did you vote against the program because it isn't worth that much money to you, or because it would be somewhat difficult for your household to pay that much, or because of some other reason?

1. **COST is too HIGH/DIFFICULT TO PAY**

   Cost is too high for R. "I can't afford it." "I don't like to pay taxes/it would be difficult for me to pay."

   Cost is too high for others. "Other people may not be able to afford it."

   Cost is too high for what it would accomplish. "The program is very expensive."

2. **NOT THAT IMPORTANT/ISN'T WORTH THAT MUCH MONEY**

Problem not that important. "Problem isn't a big deal/serious/important." "I don't care about it." "It won't help me."

Other problems are more important. "Rather spend money on other programs."

Let nature solve it. "Let nature take its own course." "Wildlife will recover anyway."

3. **CONCERNS about the program or payment plans**

Program does not do enough. "Very few animals would be helped."

Doubts about aspects of the program. "Program won't work." "Don't trust the State of California/Government." "Money will be wasted."

Concerns about payment plan. "It will cost more than you are telling me." "It (the tax) will be for more than one year."

Problem could/should be solved in other ways. (*e.g.*, making the tankers go farther offshore; getting at the root cause; etc.)

Other parties should pay. "State should use other money to pay for it." "Oil companies should pay."

Program an example of over-regulation.

4. **Respondent wants MORE INFORMATION to make a decision**

"Too many unanswered questions." "I don't have enough data."

5. **OTHER (INCLUDING NOT SURE)**

B-5 Could you tell me why you aren't sure about how you would vote?

1. **COST is too HIGH/DIFFICULT TO PAY**

Cost is too high for R. "I can't afford it." "I don't like to pay taxes/it would be difficult for me to pay."

Cost is too high for others. "Other people may not be able to afford it."

Cost is too high for what it would accomplish. "The program is very expensive."

2. **NOT THAT IMPORTANT/ISN'T WORTH THAT MUCH MONEY**

   Problem not that important. "Problem isn't a big deal/serious/important." "I don't care about it." "It won't help me."

   Other problems are more important. "Rather spend money on other programs."

   Let nature solve it. "Let nature take its own course." "Wildlife will recover anyway."

3. **CONCERNS about the program or payment plans**

   Program does not do enough. "Very few animals would be helped."

   Doubts about aspects of the program. "Program won't work." "Don't trust the State of California/Government." "Money will be wasted."

   Concerns about payment plan. "It will cost more than you are telling me." "It (the tax) will be for more than one year."

   Problem could/should be solved in other ways. (*e.g.*, making the tankers go farther offshore; getting at the root cause; etc.)

   Other parties should pay. "State should use other money to pay for it." "Oil companies should pay."

   Program an example of over-regulation.

4. **Respondent wants MORE INFORMATION to make a decision**

   "Too many unanswered questions." "I don't have enough data."

5. **OTHER (INCLUDING NOT SURE)**

APPENDIX F
Description of the Non-Parametric Turnbull Estimator[1]

1. Introduction

For this study, the Turnbull (1976) non-parametric, maximum likelihood (ML) estimator for interval-censored data is used as the summary statistic for the unobserved (or latent) variable representing the respondent's willingness to pay (WTP) for the prevention program.[2] Section 2 presents a non-technical description of the lower bound on the sample mean WTP based on the Turnbull estimator (WTP_{TL}); and its calculation is illustrated using data on the **B1CH** choice measure presented in Chapter 6. Section 3 presents an intuitive explanation of how the Turnbull estimate of the lower bound on the sample mean can be decomposed across subgroups. Section 4 provides a formal mathematical derivation of the estimator. Section 5 looks at statistical tests based upon the Turnbull estimator which can be used to examine differences between two latent variables.

2. Non-technical Description of the Turnbull Estimate of the Lower Bound

Each respondent's choice can be used to construct an interval estimate for the latent willingness-to-pay amount implied by that choice. An individual's vote (*i.e.*, choice) at a specific dollar amount will distinguish either a lower or an upper bound for his or her WTP (*e.g.*, voting *for* at the tax amount of \$25 (B1AMT) implies that the latent WTP lies in the interval \$25–∞ , and voting against at that amount implies that the latent WTP lies in the interval 0–\$25). If a respondent votes *for*, that respondent's willingness to pay (WTP) for the program is bounded from below by the tax amount B1AMT (*i.e.*, the respondent is willing to pay at least B1AMT). How much more that respondent might be willing to pay is not revealed; we know only that the respondent's WTP is not less than B1AMT (WTP $\geq$ B1AMT). If the respondent votes *not-for*, that

1 Paul Ruud played the leading role in the preparation of the material in this appendix.

2 The Turnbull framework generalizes the Ayer *et al.* (1955) model for binary discrete choice data to allow for a variety of possible patterns of censoring and truncation. The first application of the Ayer *et al.* model to CV data appears in Kriström (1990). For the first application of the more general Turnbull framework to CV data, see Carson and Steinberg (1990); and, for recent applications, see Carson, Hanemann *et al.* (1994) and Carson, Wilks, and Imber (1994). Haab and McConnell (1997; 2002) provide an interesting exposition of some of the properties of the Turnbull lower-bound mean.

respondent's willingness to pay is bounded from above by B1AMT (*i.e.*, the respondent may be willing to pay some tax amount below B1AMT or may not be willing to pay anything at all). Thus, we know only that the respondent's WTP is less than B1AMT ($0 \leq \text{WTP} < \text{B1AMT}$).[3]

Based on only this knowledge, an estimator of the sample mean proposed by Harrison and Kriström (1995), which we label WTP_{HK}, may be defined by assuming that every respondent voting *not-for* has a zero WTP and every respondent voting *for* has a WTP that is at most the particular B1AMT_i asked about. WTP_{HK} is found by summing the B1AMT_i received by respondents who voted *for* and dividing that sum by the total number of respondents (both those giving *for* and *not-for* responses).

The WTP_{HK} estimator has certain undesirable properties. First, it is an inefficient estimator because it does not make full use of the information in the observed choice measure. In particular, it does not utilize the fact that respondents were randomly assigned to different B1AMT_i. Second, and more worrisome, the estimator is inconsistent; that is, as the sample size and the number of distinct B1AMT_i used become arbitrarily large, WTP_{HK} does not converge to the population mean WTP. Indeed the WTP_{HK} estimate from a very large sample using a very large number of distinct B1AMT_i may be much further from the sample mean than the WTP_{HK} from a small sample with only one distinct B1AMT_i. Third, while both WTP_{HK} and WTP_{TL}, the Turnbull estimate of the lower bound on the sample mean, can always be shown to be less than or equal to the sample mean, WTP_{TL} is always at least as close to the sample mean as WTP_{HK}. If more than one distinct B1AMT_i is used, the WTP_{TL} will always be closer to the sample mean WTP except in a few extreme cases where WTP_{TL} and WTP_{HK} will be equal.[4] Given that both estimators are less than the desired estimate – the sample mean WTP – one would prefer the estimator that is closest to it. The downward bias, in the statistical sense of the word,[5] of the WTP_{TL} estimator with respect to the sample mean is always less than or equal to the downward bias of the WTP_{HK} estimator.

The Turnbull estimator upon which WTP_{TL} is based provides an estimate of the fraction of the population who would vote *for* at each of the distinct B1AMT_i used. The Turnbull approach relies upon the random assignment of respondents to particular B1AMT_i to provide additional information for interpreting responses at the other B1AMT_i. Table 1 displays the **B1CH** choice measure by B1AMT_i; and Table 2 reports the Turnbull estimate of the lower

[3] We have assumed that no respondent has a negative WTP (*i.e.*, no one would demand to paid compensation in return for preventing oil spills along the Central Coast); hence, the lower bound of the first interval is necessarily zero.

[4] The two measures will be equal if for all but possibly one of the B1AMT_i, 0% or 100% of the respondents vote *for*.

[5] An estimator is unbiased if, on average (in repeated samples), it is equal to the desired quantity. An estimator is biased downward if it is systematically lower than the desired quantity and biased upward if systematically higher.

Table 1. **B1CH** Choice Measure by B1AMT

B1AMT	For	Not For
$5	68.9%	31.1%
$25	56.9%	43.1%
$65	48.5%	51.5%
$120	40.3%	59.7%
$220	28.9%	71.1%
	$X^2_{(4)} = 82.48$; $p < 0.001$	

bound on the sample mean for the WTP distribution using the **B1CH** choice measure (see Tables 6.2 and 6.3 in Chapter 6). Note that the third column in Table 2 (labeled "Probability of Voting *For* at Upper-Bound") is simply the estimated fraction of those in Table 1 who would vote *for* the program at each B1AMT with the exception of the last row, [$220–∞], where the assumption that no respondent has an infinite WTP has been imposed. This fraction is estimated by taking the fractions *for* and *not-for* at a given B1AMT_i and further interpreting those fractions on the basis of the fractions *for* and *not-for* at the other B1AMT_i. If the percentage willing to pay each particular B1AMT_i always drops as the B1AMT_i increases, the Turnbull estimator will always exactly reproduce the Table 1 percentage "For" column.[6]

Table 2. Turnbull Estimate of WTP Distribution and Lower Bound on Sample Mean: **B1CH** Choice Measure [N = 1,085]

(1) Lower Bound of Interval	(2) Upper Bound of Interval	(3) Probability of Voting *For* at Upper Bound	(4) Change in Density
$0	$5	0.689	0.311
$5	$25	0.569	0.120
$25	$65	0.485	0.084
$65	$120	0.403	0.082
$120	$220	0.289	0.114
$220	∞	0.000	0.289
	Log-Likelihood	−709.48	
	Estimate of lower bound on mean	$85.39	
	Standard error of the estimate	$3.90	

6 Economic theory predicts that the percentage of the population willing to pay a specific B1AMT_i should not increase as B1AMT_i increases. Due to natural sampling variation, this result need not hold in any particular sample of the population, although one would expect violations of the theoretical prediction to decline as sample size increases and the distance between the two B1AMT_i compared increases. When such violations occur or when data with more complex censor or truncation mechanisms are used, the Turnbull estimator uses *self-consistency* principles to derive the estimates in the third column. A complete characterization is provided in Section 4.

The WTP_{TL} estimate is based on estimates of the change in density in the fourth column of Table 2. These estimates are a simple translation of the estimates in the third column of Table 2. To see this, note that the estimate of 0.311 in the fourth column is 1.00 (*i.e.*, 100% are assumed to be willing to pay $0) minus the estimate of 0.689 in the third column, the percentage who indicate they are willing to pay $5. Thus, the Turnbull estimator indicates that 31.1 percent are willing to pay between $0 and $5. In the second row, fourth column, 0.120 is found by subtracting 0.569 (the estimate from the third column) from 0.689 (the estimate from the third column, first row). The estimate of the percentage of the sample who are willing to pay between $5 and $25 dollars is 12.0 percent. The remaining entries in the fourth column are calculated similarly.

The WTP_{TL} estimate can be found by multiplying the lower bound of each interval by the fraction of the sample estimated to lie in each interval and summing the resulting column of numbers. For example, using the data in Table 2, the lower bound of the interval (displayed in the first column) is multiplied by the estimate of the fraction of the sample in each interval (displayed in the fourth column) and then the products are summed:

$$(\$0*0.311)+(\$5*0.120)+(\$25*0.084)+(\$65*0.082)$$
$$+(\$120*0.114)+(\$220*0.289)$$
$$=\$0+\$0.60+\$2.10+\$5.33+\$13.68+\$63.58=\$85.39.$$

This calculation can also be used to show the divergence between WTP_{TL} and WTP_{HK}. The WTP_{HK} estimator assumes that there is a zero probability that a respondent who votes *not-for* at $25 would pay any amount greater than $0. Using the knowledge of equivalent subsamples at the $B1AMT_i$, we estimated in Table 1 that 68.9 percent of the population are willing to pay $5 and that 56.9 percent are willing to pay $25. The difference between these two percentages, 12 percent, is the estimate of the percentage of the sample who are willing to pay at least $5 but less than $25. To assign this fraction of the sample an estimate of $0 WTP as the WTP_{HK} does is a waste of information. The WTP_{TL} estimator (see above) assumes that this 12 percent is willing to pay $5, which is the smallest monetary amount consistent with the observed data.

By effectively assuming that the estimated fraction of the sample in each interval (*i.e.*, [$0–$5], [$5–$25], [$25–$65], [$120–$220], and [$220–∞]) is willing to pay only the amount represented by the lower end of the interval, the WTP_{TL} estimator can never over-estimate the sample mean of the latent distribution. If the average willingness to pay of the fraction of the sample estimated to have a WTP amount in any interval, *e.g.*, $5 to $25, is higher than the lower endpoint of that interval, $5, then the WTP_{TL} will be lower than the desired estimate, the sample mean WTP.

From this illustrative construction of the WTP_{TL}, one can see the dependence on the particular $B1AMT_i$ used in the survey. While it is impossible to select a set of $B1AMT_i$ so that the WTP_{TL} estimate is greater than the sample mean WTP, it is possible, however, to select the $B1AMT_i$ to yield a WTP_{TL} estimate that is essentially equal to zero. The first of two straightforward ways to do this is to select a set of $B1AMT_i$ that are all very small. For example, assume the five $B1AMT_i$ used in this study had been chosen to be $0.01, $0.02, $0.03, $0.04, $0.05. The maximum possible WTP_{TL} estimate is now $0.05 which occurs if 100 percent of the respondents at each of the five $B1AMT_i$ amounts votes *for*. The other way to ensure a very small WTP_{TL} estimate is to use a set of $B1AMT_i$ that are all very large. Here, it is possible to make the WTP_{TL} estimate equal to exactly zero by choosing the set of $B1AMT_i$ so large that no respondent would be willing to pay any of the $B1AMT_i$ used.

A deeper understanding of how the WTP_{TL} estimate depends on the set of $B1AMT_i$ used may be gained by considering the following. Recall the calculation of the WTP_{TL} estimate from the previous page.

$$(\$0*0.311)+(\$5*0.120)+(\$25*0.084)+(\$65*0.082)$$
$$+(\$120*0.114)+(\$220*0.289)$$
$$=\$0+\$0.60+\$2.10+\$5.33+\$13.68+\$63.58=\$85.39.$$

A common misconception is that a very large fraction ($63.58) of the WTP_{TL} estimate ($85.39) constructed above comes from the $220 $B1AMT_i$ design point. Disregard the subsample that received the $220 design point. The WTP_{TL} estimate based only on the $5, $25, $65, and $120 design points is $56.39: ($0∗0.311) + ($5∗0.120) + ($25∗0.084) + ($65∗0.082) + ($120∗0.360). Note that this estimate of $56.39 is considerably larger than $19.81 ($85.39 minus $65.58). The fraction of the sample estimated to have a WTP amount of at least $220 is not multiplied by $0 to yield $19.81, but rather has been added to the fraction of the sample previously estimated to fall in the [$120–$220] interval and then multiplied by the lower dollar end point ($120) of the new highest interval. Thus, the increase in WTP_{TL} from having the $220 subsample is the fraction estimated to have a WTP of at least $220 times the difference between $220 and the next highest design point, $120; 0.289∗$100 = $28.90. Increasing the highest $B1AMT_i$, with the other design points fixed, may actually decrease the estimated WTP_{TL}. This is easiest to see by considering the extreme case, use of a $B1AMT_i$ so large that no one is willing to pay the amount. In this case the largest design point adds $0 to the WTP_{TL} estimate. In general, increasing the largest $B1AMT_i$ moves the WTP_{TL} estimate toward the sample mean only if the difference between the two highest design points multiplied by the fraction of the sample estimated to have a WTP greater than or equal to the highest $B1AMT_i$ used is increasing as well.

In contrast to simply increasing the largest $B1AMT_i$ which may or may not reduce the downward bias which occurs from using WTP_{TL} as an estimate of the sample mean WTP, adding an additional $B1AMT_i$ cannot decrease the

WTP_{TL} estimate in sufficiently large samples and is likely to increase it as long as the amount chosen is neither too high nor too low nor too close to another $B1AMT_i$. Indeed, as noted earlier, for sufficiently large samples, using a sufficiently large number of $B1AMT_i$, the downward bias of the WTP_{TL} becomes arbitrarily small. However, it is not typically desirable to use a very large number of $B1AMT_i$ to minimize the downward bias of the WTP_{TL} if one is also concerned about the confidence interval around the WTP_{TL} estimate. As one allocates a sample of fixed size between a larger and larger number of $B1AMT_i$ the precision with which one estimates the percentage *for* at any particular $B1AMT_i$ falls, often dramatically so, due to increased sampling variability. This increase will be reflected in an increase in the confidence interval around the WTP_{TL} estimate.

A Turnbull upper-bound mean, WTP_{TU}, may also be defined. For this estimator the fraction of the respondents estimated to be in an interval is treated as having a latent WTP at the high rather than the low end-point of the interval. The sample mean must lie between the WTP_{TL} and WTP_{TU}. WTP_{TU} is potentially infinite unless one imposes a finite upper bound on the last interval as would be suggested by the income/wealth constraint on any WTP measure.

Thus any estimate of the sample mean that is lower than WTP_{TL} or higher than WTP_{TU} is inconsistent with the observed choices made by respondents. Furthermore, without additional statistical assumptions about the nature of the latent willingness-to-pay distribution, any observed choice measure is uninformative about where, within the two Turnbull bounds, the sample mean lies. The most conservative assumption about the shape of the latent WTP distribution consistent with the observed respondent choices is that the sample mean is equal to WTP_{TL}.

3. Decomposing the Turnbull Estimate of the Lower Bound

Turnbull estimates of the lower bound for two subgroups of the sample can be compared. Two such comparisons are that of environmentalists versus non-environmentalists[7] and that of setting the votes *for* to votes *not-for* for all members of one subgroup such as those not-paying state taxes. These two questions share the same underlying statistical framework: in sufficiently large samples with random assignment of respondents to the different $B1AMT_i$, the sample Turnbull estimate of the lower bound on the sample mean can be shown to be decomposable in the sense that it represents a weighted average of the Turnbull estimates of the lower bound on the sample means of the two subgroups where the weights are the proportion of the total sample each subgroup comprises.

[7] See Section 6.5 of Chapter 6 for examples of this type of decomposition of the estimate.

Table 3. Turnbull Estimates of WTP Distributions and Lower-Bound Estimates: **B1CH** Choice Measures for Taxpayer and Non-Taxpayer Samples

		Taxpayers [N = 977]		Non-Taxpayers [N = 108]	
(1) Lower Bound of Interval	(2) Upper Bound of Interval	(3) Probability of Voting *For* at Upper Bound	(4) Change in Density	(5) Probability of Voting *For* at Upper Bound	(6) Change in Density
\$0	\$5	0.680	0.320	0	0
\$5	\$25	0.574	0.106	0	0
\$25	\$65	0.496	0.079	0	0
\$65	\$120	0.400	0.096	0	0
\$120	\$220	0.287	0.113	0	0
\$220	∞	0.000	0.287	0	0
		Estimated Lower Bound on Sample Mean: \$85.42		Estimated Lower Bound on Sample Mean: \$0	

We illustrate how the decomposition approach works by showing how the Turnbull estimate of the lower bound on the sample mean changes in going from the **B1CH** choice measure to the **B1CHNT** choice measure in Section 6.8 which sets the votes of all non-taxpayers to *not-for*.

Table 3 reports the Turnbull estimate of the lower bound on the sample means for the WTP distribution for the taxpayer sample (N = 977) and the non-taxpayer sample (N = 108) using the **B1CH** choice measure and recoding non-taxpayer *for* votes to *not for* votes. The lower-bound estimate for the taxpayer sample, \$85.42, is calculated by multiplying the lower bound of the interval (column 1) by the change in density (column 4) and by taking the sum of the products: (\$0 ∗ 0.320) + (\$5 ∗ 0.106) + (\$25 ∗ 0.079) + (\$65 ∗ 0.096) + (\$120 ∗ 0.113) + (\$220 ∗ 0.287). The estimate for the non-taxpayers in the sample is calculated in the same manner. Note that as the votes of non-taxpayers who voted *for* were recoded as votes *not for*, the probability of voting *for* (column 5) at each interval is necessarily zero; as the change in density (column 6) is simply the difference in the probability of voting *for* at each successive interval, the change in density for each interval is also zero. As shown in Table 3, the adjustment for non-taxpaying status effectively sets the lower-bound estimate for this group of respondents to zero.

The lower-bound estimate for the full sample is calculated by multiplying the estimated lower bound for each group of respondents by the percentage of the sample that the group encompasses: [(\$85.42*977) + (\$0 ∗ 108)]/1085 = \$76.92.

Table 4 reports the Turnbull estimate of the lower bound on the sample mean for the **B1CHNT** choice measure presented in section 6.8. The decomposition approach described above yields a slightly different estimate (\$76.92) than that shown in Table 4 (\$76.45) because it assumes that the percentage of the

Table 4.
Turnbull Estimate of WTP Distribution and Lower-Bound Estimate:
B1CHNT Choice Measure [N = 1,085]

(1) Lower Bound of Interval	(2) Upper Bound of Interval	(3) Probability of Voting *For* at Upper Bound	(4) Change in Density
$0	$5	0.621	0.379
$5	$25	0.519	0.102
$25	$65	0.453	0.066
$65	$120	0.354	0.099
$120	$220	0.255	0.099
$220	∞	0.000	0.255
	Log-Likelihood	−707.73	
	Estimate of lower bound on mean	$76.45	
	Standard error of the estimate	$3.78	

sample who are non-taxpayers at each design point is exactly equal when in fact, because of sampling variability, the percentages are slightly different.

4. Mathematical Derivation of the Turnbull Estimator[8]

4.1. *Basic Notation*

The general structure of the basic econometric model for survey responses has several components. The first component is latent variable y^* representing the respondent's willingness to pay for the good. The *observed* response to a survey question is the second component. We will denote this response by y. This response is determined by the latent WTP assumed to be implied by a respondent's choices, according to the nature of the survey question. For example, the simplest question asks the respondent whether he or she would vote for a program that will cost the respondent ten dollars and the respondent answers with a *for* or *not for* answer. The relationship between y^* *and* y is the third component of the basic model. In such models, this relationship is often called the *observation rule* and we will denote it by the function $\tau(\cdot): y = \tau(y^*)$.

4.1.1. *The Latent Variable*

Many of the differences among the models described here (and elsewhere) are differences in the probability distributions specified for y^*. Given an observation rule τ, the distribution of y is completely determined by the distribution one

[8] This section was taken from Carson, Hanemann *et al.* (1994), and is included here with only minor modifications.

specifies for y^*. Returning to the example, the survey question suggests the simple observation rule

$$\tau(y^*) = \begin{Bmatrix} 0 & \text{if } y^* < 10 \\ 1 & \text{if } y^* \geq 10 \end{Bmatrix},$$

where the value $y = 1$ denotes a *for* response and $y = 0$ denotes a *not for* response. If we specify that y^* has the cumulative distribution function (c.d.f.) $F(y^*)$, then the probability density function (p.d.f.) of y is the binomial

$$\Pr\{y = Y\} = F(10)^y[1 - F(10)]^{1-y}, \quad Y = 0, 1.$$

When analyzed as the marginal distribution of y^*, one can estimate the function F at points where the survey asks respondents to make a choice. The value \$10 in our example is such a point. When the y^* are analyzed conditional on explanatory variables, it is standard practice to use parametric models that specify a parametric c.d.f. for y^*. For example, F is frequently specified to be normal (Gaussian) with mean μ and scale σ,

$$F(z) = \int_{-\infty}^{z} \frac{1}{\sigma}\phi\left(\frac{s-\mu}{\sigma}\right) ds \quad \text{where } \phi(z) = \frac{1}{\sqrt{2\pi}} \exp\left(-\frac{z^2}{2}\right)$$

and then μ, and possibly σ, are specified as functions of the explanatory variables. Nonparametric models do not restrict the function F to a particular parametric family. In either case, one can use the method of maximum likelihood to derive and compute estimators for the unknown parameters.

4.2. *Observation Rules*

In this section, we describe the likelihood functions for various formats of a valuation question. These functions are the basis for maximum likelihood estimation of the parameters of the models.

4.2.1. *Single-Bound*

The simplest observation rule establishes a bound, either an upper bound or a lower bound, on the latent y^*. This is the example we used above. The bound, sometimes referred to as a *cut*, is usually an observed variable. The researcher chooses different bound values in order to identify the cumulative distribution function (c.d.f.) $F(y^*)$ at various points of its support. Denoting the bounding variable by c, the observation rule for single-bound data is

$$\tau(y^*) = \begin{Bmatrix} 0 & \text{if } y^* < c \\ 1 & \text{if } y^* \geq c \end{Bmatrix}.$$

The log-likelihood for a single observation is, therefore,

$$L(F; y) = y \cdot \ln F(c) + (1 - y) \cdot \ln[1 - F(c)],$$

because

$$\Pr\{y=0\} = \Pr\{y^* \leq c\}.$$

4.2.2. *Double-Bound*

In single-bound data, the experimenter determines two intervals, $(-\infty, c)$ and $[c, \infty)$, and observes which interval contains y^*. The double-bound observation rule establishes three intervals into which latent y^* might fall: $(-\infty, c_1)$, $[c_1, c_2)$, and $[c_2, \infty)$. Thus, the observation rule for double-bound data can be written as

$$\tau(y^*) = \left\{\begin{matrix} 1 & \text{if } y^* < c_1 \\ 2 & \text{if } y^* \leq y^* < c_2 \\ 3 & \text{if } c_2 \leq y^* \end{matrix}\right\}.$$

The log-likelihood for a single observation is, therefore,

$$L(F; y) = 1\{y=1\} \cdot \ln F(c_1) + 1\{y=2\} \cdot \ln[F(c_2) - F(c_1)]$$
$$+ 1\{y=3\} \cdot \ln[1 - F(c_2)],$$

where $1\{\cdot\}$ denotes the indicator function.

We can write this log-likelihood more compactly if we change our notation. Let $c_1(y)$ be the (possibly infinite) lower bound observed for y^* and let $c_2(y)$ be the (possibly infinite) upper bound. Then

$$\Pr\{y = Y\} = F[c_2(Y)] - F(c_1(y)]$$

and

$$L(F; y) = \ln[F(c_2(y)) - F(c_1(y))],$$

where it is understood that $F(-\infty) = 0$, and $F(\infty) = 1$.

4.3. *Turnbull Likelihood Function*

The set of indicator variables for the latent variable y^* can be analyzed as independent and identically distributed random variables. Given all the interval boundary values, upper and lower, all that one can estimate without making additional assumptions about the distribution of y^* is the probabilities of it falling between the observed boundary values defining the indicator variables. Let

$$0 \equiv c_{(0)} < c_{(1)} < c_{(2)} < \ldots < c_{(M)} < c_{(M+1)} \equiv \infty$$

be *ordered* values of the M observed finite boundaries.[9] Then the likelihood function of the observed y depends only on the M parameters

$$0 < F(c_{(1)}) < F(c_{(2)}) < \ldots < F(c_{(M)}) < 1.$$

For example, the probability (likelihood) that y^* falls in between the boundary values c_1 and c_2 is

$$\Pr\{c_1 < y \leq c_2\} = F(c_2) - F(c_1),$$

assuming that $c_1 < c_2$.

Given the observed pairs, upper and lower, of boundaries in any data set, a useful reparameterization of the likelihood function in terms of the probabilities of each of the mutually exclusive intervals determined by the $c_{(m)}$'s; that is,

$$p_1 = F(c_{(1)}), \quad p_2 = F(c_{(2)}) - F(c_{(1)}), \ldots, p_M = F(c_{(M)}) - F(c_{(M-1)}).$$

We can always write the probability of an particular interval, say $[c_1, c_2]$, in terms of the probabilities of the sub-intervals it contains. The analytical expression is

$$\begin{aligned}\Pr\{c_1 < y^* \leq c_2\} &= \sum_{m=1}^{M+1} 1\{c_1 < c_{(m)}\} \cdot 1\{c_2 \geq c_{(m)}\} \cdot p_m \\ &= \sum_{m=1}^{M} d_m p_m + d_{M+1},\end{aligned}$$

where $c_{(0)} = 0$, $c_{(M+1)} = \infty$,

$$d_m \equiv 1\{c_1 < c_{(m)}\} \cdot 1\{c_2 \geq c_{(m)}\} - 1\{c_1 < c_{(M+1)}\} \cdot 1\{c_2 \geq c_{M+1}\}, \quad m = 1, \ldots, M,$$

$$d_{M+1} \equiv 1\{c_1 < c_{(M+1)}\} \cdot 1\{c_2 \geq c_{(M+1)}\},$$

since

$$p_{M+1} = 1 - \sum_{m=1}^{M} p_m.$$

Therefore, given a vector of dummy variables $d = [d_m; m = 1, \ldots, M]$ the log-likelihood function for an observation can be written

$$L = \ln(d'p + d_{M+1})$$

as a function of the linear index $d'p$. This specification embodies the Turnbull model in which F is not restricted to belong to parametric families such as the normal family of distributions. Given a fixed number of cut points, the specification remains parametric in the sense that the parameter vector p is finite-dimensional.

[9] The parentheses around the subscripts denote an ordering. This is a common notation for order statistics, and we use it here in an analogous way. Note that in many applications it is possible that $c_{(0)} = -\infty$; we define $c_{(0)} = 0$ because the willingness to pay values have been assumed to be non-negative.

Note that every element of p must be positive and less than or equal to one for the implied probabilities to fall in the unit interval. This constraint is always satisfied by the MLE because near zero probabilities yield log-likelihood values approaching negative infinity.

Estimation of confidence intervals and statistical inference for the Turnbull estimator follow standard lines for maximum likelihood methods applied to parametric models. The limiting distribution of the maximum likelihood estimator will be multivariate normal. Based on this limiting distribution, one estimates the variance matrix for the p_m parameters using the information matrix, which is

$$E\left(\frac{\partial^2 L}{\partial p\, \partial p'}\right) = E\left[\frac{d'd}{(d'p + d_{M+1})}\right].$$

An estimator for the covariance matrix of the maximum likelihood estimator for p is

$$\hat{\mathrm{V}}\mathrm{ar}(\hat{p}) = \left[\sum_{n=1}^{N} \frac{d_n d_{n'}}{(d_{n'}\hat{p} + d_{M+1,n})^2}\right]^{-1},$$

where $\hat{p}$ denotes the estimate of p and N observations on d are indexed by n.

4.3.1 *Estimation of the Lower Bound on the Expectation*

Because the c.d.f. of y^* is identifiable at the boundary points $0 < c_{(1)} < c_{(2)} < \ldots < c_{(M)} < \infty$, a lower bound on the expectation of y^* is also identifiable. In our notation, the expectation of y^* is:

$$\begin{aligned} E(y^*) &= \int_0^{\infty} y^*\, dF(y^*) \\ &= \sum_{m=1}^{M+1} \int_{c_{(m-1)}}^{c_{(m)}} y^*\, dF(y^*). \end{aligned}$$

This moment of the distribution of y^* is not identifiable because the c.d.f. $F(y^*)$ is not identifiable at every point in the support of y^*. If one replaces y^* in the integrals above by the finite lower value of each interval,[10] one constructs a lower bound on $\mathrm{E}(y^*)$:

$$\mu_L(c) = \sum_{m=1}^{M+1} c_{(m-1)} \int_{c_{(m-1)}}^{c_{(m)}} dF(y^*) \le E(y^*).$$

This moment is identifiable because $\mu_L(c)$ is a function of the identifiable points

[10] All of the lower bounds are finite because the support of y^* is bounded below by zero in this discussion. In general, the Turnbull estimator can be applied to distributions over the real line without finite lower bounds. In such cases, this lower bound on the mean is not identifiable.

on $F(y^*)$:

$$\begin{aligned}\mu_L(c) &= \sum_{m=1}^{M+1} (c_{m-1}) \int_{c_{(m-1)}}^{c_{(m)}} dF(y^*)\\ &= \sum_{m=1}^{M+1} c_{(m-1)}[F(c_m) - F(c_{(m-1)})]\\ &= \sum_{m=1}^{M+1} c_{(m-1)} p_m\\ &= c_{L'} p,\end{aligned}$$

where we denote the vector of lower-bound values $[c_{(m)}; m = 0, \ldots, M]$ by c_L. Using the maximum likelihood estimator for p, $\hat{p}$, the estimator for $\mu_L(c)$ is $\hat{\mu}_L(c) = c_{L'} \hat{p}$. The variance of this estimator can be estimated using the estimator for $\text{Var}(\hat{p})$ given above:

$$\begin{aligned}\hat{\text{V}}\text{ar}[\hat{\mu}_L(c)] &= c_{L'} \hat{\text{V}}\text{ar}(\hat{p}) c_L\\ &= c_{L'} \left[\left(\sum_{n=1}^{N} \frac{d_n d_{n'}}{(d_{n'}\hat{p} + d_{M+1,n})^2} \right)^{-1} \right] c_L.\end{aligned}$$

If an upper bound on the support of the distribution of y^* is available, then a similar estimator for an upper bound on the expectation can be computed.

5. Tests Based on the Turnbull Estimator

5.1 *t-tests*

Estimators for $\mu_L(c)$ from two data sets can be used to test whether the underlying distributions are identical. Given random sampling, the analysis follows conventional lines. Under the null hypothesis that the distributions are the same, the statistic

$$\frac{\hat{\mu}_{L,1}(c) - \hat{\mu}_{L,2}(c)}{\sqrt{\hat{\text{V}}\text{ar}[\hat{\mu}_{L,1}(c)] + \hat{\text{V}}\text{ar}[\hat{\mu}_{L,2}(c)]}}$$

converges in distribution to a standard normal random variable as the sample size approaches infinity. One uses critical values from this distribution to construct one-sided or two-sided tests of the null hypothesis.

5.2. *Likelihood Ratio Tests*

One can also compute a likelihood ratio test of the equality of c.d.f.'s. Two distribution functions can possess the same $\mu_L(c)$ while their values for p differ. The likelihood ratio test is a test for whether all of the elements of p are the same in two populations. The test can be computed by calculating the maximized log-likelihood function for each sample, L_1 and L_2, and for the combined sample, L_C. Then $LRT = 2(L_1 + L_2 - L_C)$ has a limiting chi-square distribution under the null hypothesis that all the elements of p are equal in both sampling populations. The degrees of freedom equal the number of elements in p, which we have denoted by M.

This test may not be as powerful as the lower-bound test even though it tests a more restrictive null hypothesis. The lower-bound test has fewer degrees of freedom and can be a one-sided test. Both of these factors will contribute to its power relative to the likelihood ratio test.

APPENDIX H
Supplemental Analysis Tables

1. Introduction

This appendix presents supplemental tables referenced in Chapter 6. Section 2 presents an alternative specification for the multivariate choice function estimated using a Weibull model with **B1CH** as the dependent indicator variable and with the covariates included in the probit model (see Table 6.7) as independent variables. In section 3, the variable definitions and parameter estimates used in the income imputation equation are presented. Section 4 provides an estimate of the probit model without the observations that are missing income.

2. Weibull Valuation Function

Table 1. Weibull Estimates for **B1CH** Choice Valuation Function [N = 1,085]

Parameter	Parameter Estimate	Standard Error	Z-Statistic
LOCATION	−1.935	1.831	−1.05
SCALE	2.465	0.255	9.66
LINC1	0.405	0.170	2.39
LINC2	0.275	0.156	1.75
NOTAX	0.501	0.508	0.98
CCOAST	0.814	0.259	3.14
COASTIP	0.973	0.324	3.00
WILDIP	1.296	0.298	4.34
ENVIST	1.113	0.353	3.14
LOWSPEND	−1.109	0.511	−2.17
PAYVEH	1.378	0.297	4.63
HWY1	0.818	0.382	2.14
FAMBIRD	0.851	0.361	2.35
MOREHARM	0.480	0.298	1.61
LESSHARM	−0.892	0.357	−2.50
PMWORKS	−1.827	0.307	−5.94
PNOTWORK	−3.130	0.565	−5.53
PAYMORE	−0.545	0.259	−2.13
PROTEST	−1.614	0.332	−4.86

Log(L) = −514.72.

3. Income Imputation

In question D-12, respondents were asked to select one of 11 household income categories (or intervals) from a showcard. The lowest interval was \$0 to \$10,000 and the highest, \$150,000 or more. The mean income for each interval category from the 1990 Census for California households was used as the income value for those respondents who selected an income category. In most cases, the mean estimate for each interval was fairly close to the mid-point of the interval. The two exceptions were the lowest interval (mean = \$2,402) and the open-ended, highest interval of \$150,000 or more (mean = \$212,953).

Some respondents (86 out of 1,085 or 8%) chose not to provide information regarding their household income. For those respondents, the log of household income was imputed using a regression equation that predicted the log of income from those respondents who provided this information. Table 2 provides definitions of the variables used in the equation, and Table 3 exhibits the income prediction equation. The principal predictor variable in the regression equation was the log of 1990 median household income by zipcode (as reported by the U.S. Census Bureau). Other variables included the type of dwelling unit of the household, the number of employed adults in the household, as well as the educational level, sex, race, and age of the respondent.

Table 2. Definitions for Variables Used in Income Equation

ZIPLINC equals the natural log of the 1993 median income in the respondent's zipcode; missing values replaced by the mean value.
SFAM equals one if the respondent resides in a single-family home (SPS Screener) and zero otherwise.
THOUSE equals one if the respondent resides in a townhouse (SPS Screener) and zero otherwise.
MOBILE equals one if the respondent resides in a mobile home (SPS Screener) and zero otherwise.
EMP1 equals one if the respondent indicated that exactly one adult in his/her household currently works for pay (D-11 = 1) and zero otherwise.
EMP2 equals one if the respondent indicated that exactly two adults in his/her household currently work for pay (D-11 = 2) and zero otherwise.
EMP3 equals one if the respondent indicated that exactly three adults in his/her household currently work for pay (D-11 = 3) and zero otherwise.
EMP4 equals one if the respondent indicated that exactly four adults in his/her household currently work for pay (D-11 = 4) and zero otherwise.
EMP5 equals one if the respondent indicated that exactly five adults in his/her household currently work for pay (D-11 = 5) and zero otherwise.
EMP6 equals one if the respondent indicated that six or more adults in his/her household currently work for pay (D-11 = 6) and zero otherwise.
HSCHOOL equals one if the respondent indicated that the highest educational degree he/she has received was a high school diploma (D-10 = 4) and zero otherwise.
SCOLLEGE equals one if the respondent indicated that the highest level of education he/she has completed was some college but not the equivalent of an associates degree in an academic program (D-10 = 5) or that he/she had earned a diploma in an occupational or vocational program (D-10 = 6) and zero otherwise.
COLAAD equals one if the respondent indicated that the highest educational degree received was an associates degree from an academic program (D-10 = 7) and zero otherwise.
COLGRAD equals one if the respondent indicated that the highest educational degree he/she received was a bachelor's degree (D-10 = 8) and zero otherwise.
MASTERS equals one if the respondent indicated that the highest educational degree he/she received was a master's degree (D-10 = 9) and zero otherwise.
PDEGREE equals one if the respondent indicated that the highest educational degree he/she received was a professional school degree, *e.g.*, M.D., J.D., (D-10 = 10) and zero otherwise.
DDEGREE equals one if the respondent indicated that the highest educational degree he/she received was a doctorate degree, *e.g.*, Ph.D, Ed.D, (D-10 = 11) and zero otherwise.
FEMALE equals one if the respondent was female (E-1 = 2) and zero otherwise.
HISPANIC equals one if the screener respondent indicated his/her household's national origin or ancestry was Mexicano/Mexican/Mexican American/Chicano, Puerto Rican, Cuban, Central/South American, or Other Spanish/Hispanic (SPS Screener) and zero otherwise.
BLACK equals one if the screener respondent indicated his/her household's racial background is Black/African American (SPS Screener) and zero otherwise.
ASIAN equals one if the screener respondent indicated his/her household's racial background was Asian/Pacific Islander and zero otherwise.
AGE equals the age of the respondent (D-9); missing values were replaced by the sample mean age of 45.
AGE2 equals the square of the respondent's age (D-9).

Table 3. Log Income Prediction Equation

Variable	Coefficient	Standard Error	Variable Mean
CONSTANT	2.5138	0.9941	NA
ZIPLINC	0.5442	0.0937	10.5565
SFAM	0.2886	0.0617	0.6157
THOUSE	0.2168	0.1259	0.0498
MOBILE	−0.4719	0.1430	0.0396
EMP1	0.5668	0.0913	0.3650
EMP2	0.9717	0.0949	0.3705
EMP3	1.1245	0.1325	0.0608
EMP4	1.2835	0.2404	0.0129
EMP5	1.7514	0.8211	0.0009
EMP6	1.3380	0.5750	0.0018
HSCHOOL	0.3054	0.0871	0.2286
SCOLLEGE	0.4687	0.0856	0.2747
COLAAD	0.6797	0.1402	0.0452
COLGRAD	0.7023	0.0942	0.2009
MASTERS	0.9287	0.1261	0.0645
PDEGREE	1.2924	0.1900	0.0212
DDEGREE	0.7737	0.2383	0.0129
FEMALE	−0.2334	0.0515	0.5060
HISPANIC	−0.1649	0.0772	0.1521
BLACK	−0.4480	0.1074	0.0654
ASIAN	−0.2231	0.1211	0.0488
AGE	0.0400	0.0093	45.1963
AGE2	−0.0003	0.0001	2320.794

N = 999; Adjusted $R^2 = 0.413$.

4. Construct Validity Equation with Missing Income Observations Dropped

Table 4. Multivariate Analysis of Construct Validity: Probit Estimates for **B1CH** Choice Valuation Function with Missing Income Observations Dropped

Variable	Coding	Parameter Estimate	Z-Statistic	p-value (two-sided)	Variable Mean
CONSTANT	Equals 1 for all respondents	−1.5941	−2.14	0.032	NA
B1AMT	B1 tax amount	−0.1133	−1.52	0.129	87.4224
λ(B1AMT)	Box-Cox parameter	0.2731	1.27	0.204	NA
LINC1	Log of income if < \$150,000; 0 otherwise	0.1872	2.99	0.003	9.9820
LINC2	Log of income if ≥ \$150,000; 0 otherwise	0.1482	2.44	0.015	0.4667
NOTAX	Did not pay California taxes = 1; 0 otherwise	0.1560	0.89	0.373	0.1081
CCOAST	Resides in Central Coast PSU (807, 812, 813, and 814) = 1; 0 otherwise	0.1692	1.72	0.086	0.4565
COASTIP	A-1e protecting coastal areas very important or extremely important = 1; 0 otherwise	0.4912	3.74	0.000	0.7798
WILDIP	A-2c spending to protect wildlife very important or extremely important = 1; 0 otherwise	0.4882	4.49	0.000	0.5756
ENVIST	D-7 strong environmentalist or activist = 1; 0 otherwise	0.3347	2.71	0.007	0.2162
LOWSPEND	Wants spending only on one or no programs (A-2a, A-2b, A-2d, and A-2e) = 1; 0 otherwise	−0.5174	−2.26	0.024	0.0681
PAYVEH	D16 prefer tax vehicle over higher prices or indifferent = 1; 0 otherwise	0.4718	4.51	0.000	0.4104
HWY1	Traveled along the Central Coast on Highway 1 = 1; 0 otherwise	0.2912	1.95	0.051	0.8839
FAMBIRD	Familiar with any of five types of birds often harmed in oil spills = 1; 0 otherwise	0.2222	1.50	0.134	0.8699
MOREHARM	C-1 oil spills more harmful than described = 1; 0 otherwise	0.1949	1.74	0.082	0.3473
LESSHARM	C-1 oil spills less harmful than described = 1; 0 otherwise	−0.3345	−2.22	0.026	0.1572
PMWORKS	C-3 expect program to be somewhat effective = 1; 0 otherwise	−0.6632	−6.39	0.000	0.3834
PNOTWORK	C-3 expect program to be not too effective or not effective at all = 1; 0 otherwise	−1.5342	−7.01	0.000	0.0781
PAYMORE	C-4 does not think will only have to pay special tax for one year = 1; 0 otherwise	−0.2923	−2.97	0.003	0.4304
PROTEST	Stated oil companies should pay for program or that oil companies would pass program costs on to consumers = 1; 0 otherwise	−0.7289	−5.69	0.000	0.1872

N = 999, Log(L) = −463.65, Pseudo $R^2 = 0.357$.

APPENDIX I
Comparative Analysis of COS Survey and EVOS Survey

1. Introduction

As this study bears several similarities to a study of damages caused by the *Exxon Valdez* oil spill (Carson *et al.*, 1992), it may seem natural to compare the estimates from these two studies. However, neither the populations sampled, the location of the injuries, the relationship between the location of the injuries and the populations sampled, or the nature of the injuries avoided are directly comparable. In particular, the difference between the two studies in the relationships between the residences of the respondents and the location of the injuries is striking: the COS study estimates the value to *California households* of preventing spills along the *California coast*, and the *Exxon Valdez* oil spill study (EVOS) estimates the value to *U.S. households outside Alaska* of preventing a spill on the *Alaskan coast*.[1] Section 2 compares the features of the two survey instruments and sample designs.

A direct comparison of the two studies is further complicated by the presentation of different principal summary statistics in the two reports; as they are, the two estimates are not comparable at all. However, similar summary statistics for the two willingness-to-pay (WTP) distributions may be calculated. Section 3 presents estimates for the two studies of the Turnbull lower bound on the sample mean WTP and of median WTP which suggest that average WTP's for the two goods are close. However, two extremely important caveats condition this observation. First, these amounts represent two different populations valuing two different goods. Second, while the respective point estimates of average WTP for the two studies may be close, the underlying WTP distributions which produce those point estimates are very different. This difference is revealed by formally testing whether, conditional on a large number of covariates common to the two studies, the distributions of WTP responses are statistically equivalent. Section 4 presents a test of the equivalence of the parameters of the construct validity equations estimated from the two studies.

[1] The non-inclusion of Alaskan households was an artifact of the probability sample for the EVOS study. *See* Carson *et al.*, 1992.

2. Overview of EVOS and COS

a. Survey Instruments

The general framework of the EVOS and COS survey questionnaires is quite similar: each consists of a sequence of the same basic survey components and both use an escort ship program as the prevention mechanism. Although the framework is similar, the attributes of the two objects of choice described in their respective scenarios are very different. Chapters 2 and 3 describe in detail the scenario in the COS main study survey instrument. Below, we summarize the structure of the survey instrument used in the EVOS study and, within that context, the EVOS object of choice and central scenario features.[2]

The object of choice in the EVOS survey questionnaire was characterized as a program which would prevent a future *Exxon Valdez* type oil spill that scientists expect to occur in Alaska's Prince William Sound sometime over the next ten years. The context in which this choice was presented included the following scenario features: (1) a description of Prince William Sound, the transport of oil through the Sound by ship, and the *Exxon Valdez* oil spill and its effects, (2) the presentation of an escort ship program which would prevent damage from another spill that would have the same effect on the environment as the Valdez spill, and (3) the description of a payment mechanism whereby taxpayers would make a one-time federal tax payment that would go into a Prince William Sound Protection Fund.

A referendum format was used to elicit respondents' choices: respondents were asked how they would vote on the escort ship program if the program cost their household a specified tax amount. The four dollar amounts asked about were \$10, \$30, \$60, and \$120.[3] Other questions preceding and following the choice elicitation ask about respondent attitudes, awareness of the *Exxon Valdez* spill, past experience with the affected resources, understanding of the assumptions underlying the scenario,[4] and personal or household characteristics. In the course of the interview, respondents were shown nineteen visual aids consisting of maps, color photographs, and show cards.

[2] See also Carson *et al.*, 1992, and Carson *et al.*, 1997.

[3] The EVOS survey instrument used a double-bounded dichotomous-choice elicitation framework. Conditional on respondents' answers to the first vote question, a follow-up vote question asked about either a smaller tax amount (if voted *against* or *not sure* on the first choice) or a larger tax amount (if voted *for* on the first choice). The COS instrument contained only one choice question; and thus the responses to the second, follow-up question in the EVOS survey are not used in the comparative analysis presented below.

[4] Another dimension along which EVOS differs from COS is the level of acceptance of the scenario's key features. For example, 35% of the COS sample thought that the expected harm over the next ten years would be a lot more than that described whereas the comparable percentage from EVOS was 8%; and 39% of the COS sample assumed they would have to pay the special tax for more than one year whereas only 23% of the EVOS sample felt this way.

b. Sample Design

Chapter 4 discusses the administration of the COS main study; below, we summarize the execution of the EVOS study. Westat, Inc., the survey firm retained to administer the COS survey, also administered the EVOS survey. EVOS was conducted using a multi-stage area probability sample of residential dwelling units drawn from the 50 United States and the District of Columbia.[5] In the first stage of selection, 61 PSU's (Primary Sampling Units), consisting of a county or a group of counties, were drawn with probabilities proportionate to their population counts.[6] Within these selected PSU's, 334 Census blocks[7] (or block groups) were chosen with probabilities proportionate to their population counts. In the third stage, 1,554 dwelling units were randomly drawn from the enumerated listing of selected blocks. The overall response rate for EVOS was 75.2 percent.[8]

The distinguishing scenario features and the sampling frames for the EVOS and COS studies are summarized below in Table 1.

As noted above, the COS sampling frame is the state of California whereas the EVOS frame is the United States. In addition to the obvious geographical differences between the sampling frames, the California population is known to have a substantially higher proportion of environmentalists as well as higher incomes relative to the U.S. as a whole. Unfortunately, while EVOS interviewed a random sample of the U.S., the sampling design used does not allow for statistical inference about the population of California to be drawn from the California households sampled.

3. Statistical Analysis

a. Per Household Damage Estimates

The principal summary statistic reported in the COS study is the Turnbull estimate of the lower bound on the sample mean WTP; the principal summary statistic reported in the EVOS study is median WTP. The parametric assumption adopted in calculating the estimate of the mean of a random variable can

5 Because Alaska and Hawaii were excluded from Westat's original sampling list, a new stratum containing these two states was created. A random selection of PSU's from this stratum yielded the Honolulu SMSA.

6 Before this first stage was executed, the list of PSU's was stratified by eight 1980 Decennial Census characteristics: region of the country, SMSA versus non-SMSA, rate of population change between 1970 and 1980, percentage living on a farm (for non-SMSA PSU's), percentage employed in manufacturing, percentage white, percentage urban, and percentage over age 65. Selection by strata typically increases the precision of the survey results.

7 The Census blocks were stratified by two block characteristics: percentage of population that was black and a weighted average of the value of owner-occupied housing and the rent of renter-occupied housing.

8 This response rate was calculated assuming that all of the unknown eligibility cases were in fact eligible; the comparable COS response rate is 73.5 percent (see Chapter 4).

Table 1. EVOS-COS Feature Comparison

Study Feature	EVOS	COS
Object of Choice	program that would prevent another *Exxon Valdez*-type spill sometime over next 10 years	program that would prevent cumulative harm from oil spills along California's Central Coast over the next 10 years
Injury Description	1,000 miles of shoreline oiled – few years to recover; 22,600 dead birds found with estimated bird deaths of 75,000 to 150,000 – 3 to 5 years to recover; 580 otters and 100 seals killed – couple of years to recover	many small animals and plants along 10 miles of shoreline killed – 5 years to recover; 12,000 birds killed and 1,000 injured – 10 years to recover
Payment Vehicle	one-time increase in federal income taxes; money goes into Prince William Sound Fund	one time increase in State income taxes; money pays to set up response centers
Location	South Central Alaska Coast	Central California Coast
Sampling Frame	U.S. residents	CA residents

play a considerable role in determining the magnitude of that estimate. The Turnbull approach (and generally the median approach) avoids the problems involved in making parametric distributional assumptions but at the expense of obtaining an estimate which is known to be a lower bound on the desired measure. Since software for the Turnbull estimator of the lower bound on the sample mean and its standard error was not generally available at the time of EVOS, the median was selected.

Another difference between the COS and EVOS studies is that the COS study has only one choice question (exclusive of the reconsideration questions); the EVOS study had an initial choice question and a follow-up choice question. To improve comparability, only the first choice question in EVOS and the single choice question in COS are used below to estimate single-bounded Weibull medians.

Estimating the Weibull median using just the responses to the first EVOS choice question and the reconsideration question results in a single-bounded value of $38.11 with a 95 percent confidence interval of [$29.87–$48.62].[9] The

9 Previously reported EVOS willingness-to-pay estimates are the double-bounded Weibull median of $30.91 reported in Carson *et al.*, 1992 and the double-bounded Turnbull lower-bound mean of $53.60 reported in Carson *et al.*, 1997. For the Weibull median, the estimate using the single-bounded data is higher, $38.11 compared to $30.91. A downward bias is usually introduced by the follow-up choice question used for generating the double-bounded estimate. Furthermore, the additional data points tend to better define the right tail of the Weibull, driving the estimate down. For the Turnbull estimate of the lower bound on the sample mean, the estimate using the double-bounded data is higher, $53.60 compared to $51.70. The double-bounded Turnbull estimate may also sometimes be lower, as one is

Table 2. $VOTE_E$ and $VOTE_C$ by AMT

$VOTE_E$ – EVOS [N=1,043]			$VOTE_C$ – COS [N=857]		
AMT	For	Not For	AMT	For	Not For
$10	67.4%	32.6%	$5	69.0%	31.1%
$30	51.7%	48.3%	$25	56.9%	43.1%
$60	50.6%	49.6%	$65	48.6%	51.5%
$120	34.2%	65.8%	$120	40.3%	59.7%

comparable COS Weibull median based on the responses to the single choice question and the reconsideration questions administered in this study is $44.95 [$33.95–$59.51].

Such a straightforward comparison using the Turnbull lower-bound estimator is not feasible: the difference between the Turnbull estimate of the lower bound on the sample mean and the sample mean is a function of the design points.[10] In order to compare lower bounds on the sample means, use of the same design points is desirable; in this case, the first four COS design points are sufficiently close to the four EVOS design points to enable a reasonable comparison. As noted above, respondents in the EVOS study were randomly assigned to one of four versions of the questionnaire which differed only by the tax amount: $10, $30, $60, or $120. In the COS study, one of five tax amounts was randomly assigned: $5, $25, $65, $120, or $220. For comparison, we constructed choice measures using responses to the first EVOS choice question ($VOTE_E$) and the responses to the single COS choice question for the first four tax amounts ($VOTE_C$). Both choice measures also take into account responses to the reconsideration questions. Table 2 displays the percentages of *for* and *not-for* responses for $VOTE_E$ and $VOTE_C$ by the tax amount (AMT) for the two studies.[11]

Table 3 reports the Turnbull estimate of the lower bound on the sample mean for the WTP distribution using the $VOTE_E$ and $VOTE_C$ choice measures described above. The Turnbull estimate of the lower bound on the sample mean for EVOS respondents' willingness to pay for the plan to prevent another

effectively averaging two separate estimates of the percentage in favor at some tax amounts; if the percentage voting for the tax amount is lower at the followup question than it was at the first question, the followup percentage may drive the estimate lower. This effect may counter the usual statistical property of the Turnbull estimate that compels it to increase toward the sample mean from below as the number of design points increases.

10 Theoretically, the estimate can not decrease as additional design points are added; see Appendix F.

11 As in Chapter 6, *not sure* and *refused* responses were treated as *not-for* votes.

Table 3. Turnbull Estimates of the Lower Bound on the Sample Means

	EVOS [N = 1,043]	COS [N = 857]
Basis for Estimate	full sample	sample administered first 4 design points
Estimate of lower-bound mean	$51.70	$56.44
Standard error of the estimate	$2.10	$2.48

Exxon Valdez type oil spill is $51.70;[12] and, using the first four design points, the estimate of COS respondents' willingness to pay for the program to prevent injuries from oil spills along California's Central Coast is $56.44.

A formal statistical test of both the median and the Turnbull estimate of the lower bound on the sample mean shows that per household WTP in the two studies is not statistically different at the 95% confidence level. However, this finding should not be taken to indicate any equivalence of the underlying WTP distributions. The households in the two samples being compared are likely to be quite different; in particular, the EVOS estimate is based on a random sample of U.S. households while the COS study is based on a random sample of California households. This difference may be controlled to a large degree by using the set of covariates, such as demographic variables, which are available from the two studies. We now turn to this comparison.

b. Comparison of Construct Validity Equations

The following set of covariates is available for both the EVOS and COS sample: log of the bid amount (LOGAMT), log of income (LINC), variables identifying respondents who believed the oil spill(s) would cause more (MOREHARM) or less harm (LESSHARM) than that described in the survey, variables identifying those who believed the program would be only somewhat effective (PMWORKS) or not effective (PNOTWORK), a variable identifying those who believed they would have to pay the tax for the program for more than one year (PAYMORE), a variable identifying those who stated protecting coastal areas from oil spills was very or extremely important (COASTIP), a variable identifying those who consider themselves environmentalists (ENVIST), a variable identifying those who are caucasian (WHITE), and a

[12] The sampling design was stratified by region of the country (*i.e.*, each region was randomly sampled separately), so while the design does not permit statistical inference about California, it does permit statistical inference about the Pacific Coast region (which includes California as well as Washington, Oregon, Hawaii, and Alaska). The Turnbull estimate of the lower bound on the sample mean for Pacific Coast respondents sampled in EVOS is $57.86 with a standard error of $5.63. As expected, this value is somewhat higher than that for the total EVOS sample.

variable identifying those who stated that the oil companies should pay for the program costs (PROTEST).

Separately estimating a probit model for each sample allows for different coefficients for each of the 11 covariates. The log-likelihoods from these two models can be compared to that from a model which pools the EVOS and COS samples. This combined model imposes the restriction that the coefficients on the covariates in the two samples are alike. A $\chi^2(11)$ test indicates rejection of the null hypothesis that the covariates from the two samples are alike. Taking twice the difference in the sum of the two log-likelihood values for the separate models and the log-likelihood for the combined model results in $\chi^2(11) = 111.894$ which can be rejected at $p < 0.001$.[13] Further, if location-specific covariates (*e.g.*, variables measuring interest in, use of, and proximity to the affected natural resources) were available for both samples, the inclusion of these would likely result in an even larger difference between the two estimated construct validity equations.[14]

An alternative to estimating separate models for the two samples is to estimate a combined model that includes interaction terms between the covariates and a dummy variable identifying the sample (in this case, we use an EVOS dummy). An advantage of this approach is that the level of significance of the interaction covariates indicates the principal sources of divergence between the two construct validity models. In this case, those interaction terms that exhibit statistical significance are WHITE, PNOTWORK, PAYMORE, COASTIP, ENVIST, and PROTEST.

4. Concluding Remarks

EVOS values the prevention of a very large spill geographically distant from most of the population sampled whereas COS values the prevention of a series of smaller spills in relatively close proximity to the population sampled. There is no *a priori* reason the COS values should be either higher or lower than those from EVOS. While the Turnbull estimates of the lower bound on the sample means and the Weibull medians are similar for the two studies, a statistical analysis conditioning on the covariates common to both studies demonstrates that the distributions of WTP in the two studies are quite different.

[13] The EVOS survey instrument was also used in another study that was administered approximately 18 months before the administration of the COS survey (see Carson *et al.*, 1997). A comparison of those results to the original EVOS results (from two years earlier) indicated nearly identical univariate estimates of the latent WTP distribution. Further, it was not possible to reject that the coefficients in the construct validity equations were identical in the two EVOS samples.

[14] For example, the COS results suggest that the values are strongly influenced by measures of geographic proximity (*e.g.*, CCOAST and HWY1); see Chapter 6.

APPENDIX L
Reply to Triangle Economic Research Critique

Introduction

On behalf of an unspecified group of companies, Dunford *et al.* (1996) [hereafter TER] reviewed the report "The Value of Preventing Oil Spill Injuries Along California's Central Coast" [hereafter the COS Report].[1] Their review[2] [hereafter Critique] attributed a number of flaws to the COS Study and concluded (p. vi) that "[b]ecause of its serious problems, the COS Study will not be useful for assessing natural resource damages for several reasons." TER conclude (p. iv) that the COS Study has six major flaws:

(1) "The hypothetical nature of the responses",
(2) "Many respondents did not provide a WTP response for the described scenario",
(3) "The failure to test for sensitivity to the scope of the scenario",
(4) "Use of a flawed analytical approach for estimating average WTP",
(5) "[D]esign flaws in the COS study [that] preclude its use for assessing compensable value losses for oil spills that differ from the spill described in the study", and
(6) "[D]esign flaws in the COS study [that] preclude its use for scaling compensatory restoration options under the new NOAA natural resource damage assessment (NRDA) regulations".

In this reply, we refute each of TER's several claimed major flaws in the COS Study.

The assertions in the TER Critique fall into four main categories: (a) general claims that contingent valuation (CV) does not work as a method, (b) claims that the COS Study might have some problem that TER only vaguely describes and for which TER do not present any real support, (c) claims that a standard or accepted practice used in the COS Study is inappropriate, and (d) claims that the COS Study has a specific problem for which the TER Critique cites specific evidence from the COS Report. We consider each of the major flaws

1 The COS Study was reported in Carson, Conaway *et al.* (1996).

2 A copy of the TER Critique may be obtained from Triangle Economic Research (www.ter.com) or from bama.ua.edu/~issr/cosbook.html.

claimed by TER; and for a discussion of general issues in the debate over contingent valuation, we refer the reader to other papers that we have written.[3]

Major Flaw 1: Hypothetical Nature of Responses

The general nature of the TER Critique on this point is that there is an "upward" *hypothetical bias* in the responses to CV willingness to pay (WTP) questions which the COS Study shares:

> The results of these stated [WTP] versus actual [WTP] studies clearly show an upward bias (termed hypothetical bias) in CV estimates arising from the hypothetical nature of the question. [p. iv]

TER claim that CV estimates are inconsistent with estimates for the same goods obtained using other methods and that the CV estimates are higher. TER characterize this as "hypothetical bias." TER then criticize the NOAA Panel's endorsement of the referendum model as a means of avoiding some of the hypothetical bias problem. On the basis of selective data, TER further argue that hypothetical referenda may not predict real referenda. TER then note that even the NOAA Panel thought that hypothetical referenda did not ensure CV results would be economically sound. Finally, TER argue that the lack of a calibration factor to apply across all CV studies is a shortcoming of CV.

CV surveys are to some degree hypothetical in that they incorporate and rely upon facts that are chosen for the purpose of constructing a plausible scenario to use in a survey to value a particular good. However, the term *hypothetical bias* employed by TER is a misnomer. The common effect of a scenario's lack of plausibility or realism is not bias, which is directional error, but random, directionless error (Mitchell and Carson, 1989, pp 191; 217). Therefore hypothetical facts in and of themselves will not harm the validity of CV estimates (Mitchell and Carson, 1989, pp 191; 217). The Braden, Kolstad, and Miltz (1991) paper cited by TER concerns a situation not relevant to the situation posed to respondents in the COS Study – the choices their respondents make have no potential cost to them. The Smith paper (1986) TER cite

[3] There were two major symposiums on CV: one in the American Economic Association's *Journal of Economic Perspectives* to which Hanemann (1994) contributed a paper, and one in the American Agricultural Economics Association's *Choices* to which Carson, Meade, and Smith (1993) contributed a paper. Several of the authors of the TER Critique contributed a paper to the later symposium (Desvousges *et al.*, 1993). The major critique of contingent valuation is contained in the Exxon-sponsored volume edited by Hausman (1993). The Arrow *et al.*, 1993 NOAA Blue Ribbon Panel [hereafter NOAA Panel] Report contains an independent assessment of CV. More recent, general papers on CV which deal with many of the criticisms raised in the TER Critique are Carson (1997a; 1997b; 2000), Carson, Flores, and Meade (2001), Carson, Groves and Machina (1999), Carson and Mitchell (1993; 1995a, 1995b), and Hanemann (1994; 1995; 1996).

concludes that respondents need tangible incentives to respond accurately. Most modern CV surveys, like the COS survey, go to substantial lengths to convey to respondents that policymakers will take their answers into account in making decisions concerning the natural resource in question. The real issue is not that facts in a CV survey are hypothetical but whether respondents find the scenario plausible and therefore take the contingencies described in the scenario into account when they value the good by answering the willingness-to-pay (WTP) question. We have evidence the respondents to the COS survey do find its scenario plausible which we discuss below.

Regardless of the appropriateness of the term *hypothetical bias*, TER's claim that CV studies overestimate the consumer's WTP for a good or for preventing a bad needs to be addressed. To support their contention that CV studies are biased upward, TER cite several studies which compare CV estimates with actual monetary contributions for the same good.

The TER Critique (p. 4) cites several of these studies – Duffield and Patterson, (1992); Seip and Strand (1992); Kealy, Montgomery, and Dovidido (1990); and Bohm (1992; note published version is 1994) – and places particular emphasis on a recent study by Brown *et al.* (1996)[4] which looks at survey responses versus actual contributions for a program to remove old roads in the Grand Canyon. The major problem for TER's argument is that most of the comparisons they reference use voluntary contributions as the payment mechanism. According to standard microeconomics texts (*e.g.*, Varian, 1992) (and ignoring the issue of the quality of the CV survey on which a given WTP estimate is based), the amount of money collected via voluntary contributions will substantially underestimate the public's true willingness to pay for a program (the desired economic measure) due to the classic free riding problem. Thus, comparing the survey measure of WTP to a measure derived from actual contributions cannot say anything about the divergence between the survey measure of WTP and the public's true WTP, a point which Champ *et al.* (1997) stresses and which TER do not mention.

The TER review of this literature does not even mention the largest comparison of CV and behavior based estimates of WTP by Carson, Flores *et al.* (1996) in which, based on a comprehensive review of the literature, 616 comparisons of CV estimates with revealed preference based estimates derived from techniques like hedonic pricing and travel cost analysis were presented. On average, CV estimates were somewhat smaller than those based on revealed preference methods and estimates of the value of goods based on the two methods were highly correlated. The most relevant evidence for the COS payment mechanism – tax payments with a referendum vote – comes from looking at studies (*e.g.*, Carson, Hanemann, and Mitchell, 1987; Champ, 1997; Vossler *et al.*, 2003) which compare the percentage in favor of providing a public good from a CV survey conducted before an election with the percentage

[4] See also Champ *et al.* (1997).

in favor in the actual election. In instances where information on the good and its costs does not differ between the two time periods,[5] these studies do not support the TER position.

The TER Critique (p. 5) suggests that the authors of the Imber, Stevenson, and Wilks (1991) and Silberman, Gerlowski, and Williams (1992) studies say they tried to provide incentives for truthful preference revelation but admitted that they failed in their efforts. While these two articles discuss the difficulties of providing incentives for truthful preference revelations, they do not take the position that their studies failed. The TER Critique points out that the NOAA Panel notes possible reasons that CV estimates might be too high. We should point out that the NOAA Panel report also points out reasons that CV estimates might be too low. The Panel notes that from a theoretical perspective a WTP measure (the quantity that a CV survey is intended to estimate) is always less than or equal to the corresponding WTA measure, which is the correct one from a damage assessment perspective. Hanemann (1991; 1999) has shown that the difference between the WTP and WTA measures for public

[5] The TER Critique cites an assertion by Diamond and Hausman (1993) that the frequent divergences between voting intentions as expressed in polls and actual votes suggests that surveys are unreliable. TER recount a particular case purported by Diamond and Hausman to be evidence of such divergence: an early LA Times Poll indicating 55% would vote in favor of Proposition 128 in California while only 36% actually did. However, Diamond and Hausman did not seem to consider all the facts. First, agent preferences for a measure like Proposition 128 should change in the face of changing information about it. Yet Diamond and Hausman disregard the massive information campaigns (or disinformation campaigns, depending upon one's perspective and actual knowledge of what a very complex ballot measure would do and cost) being waged by both sides of this Proposition. The standard profile for a ballot initiative where the proponents and opponents spend substantial amounts to influence public opinion is for support for the measure to be at its high point soon after it qualifies (because monetary expenditures have been largely one-sided up to this point in favor of the measure, with proponents telling the truth but not necessarily the whole truth). From that point, support for the measure tends to decline as a function of the relative amount of money spent in opposition (with the opponents also telling the truth, but not necessarily the whole truth.) In the case of Proposition 128, the opponents outspent proponents by a margin of almost three to one and managed to cast the issue in part as a referendum on Tom Hayden, who had helped draft the measure and was its major financial supporter. Second, Diamond and Hausman fail to mention that in the LA Times Poll at the end of September, Hayden was negatively viewed by Californians by a margin of 5 to 3; and with Hayden's support for Proposition 128 mentioned to respondents, the percentage in favor of Proposition 128 fell to 44%. Third, Diamond and Hausman fail to mention that the LA Times Poll on October 27th, a week and a half before the election, showed Proposition 128 to be trailing by 12 percentage points among likely voters (only 2% different from the actual vote). Finally, the Field Institute conducted a split ballot experiment which found that support for Proposition 128 fell by 10% if the respondent was read the fiscal impact statement for Proposition 128's (prepared by State of California's Legislative Analyst) which would appear on the actual ballot along side a summary of the proposition. This result suggests that early survey questions on Proposition 128 that simply read the ballot summary without the fiscal impact statement would provide a upwardly biased estimate of the percentage who would vote in favor. The case of Proposition 128 thus seems to lend little support to TER's position.

goods may be quite large. As a consequence, there may be a large downward bias in damage estimates that are based on WTP measures. Further, the NOAA Panel also recommends the adoption of conservative design and estimation strategies as an added precaution to ensure that over-estimation does not occur.

The TER Critique contends that there are four key differences between referendums and CV surveys that raise problems for the use of estimates based upon CV surveys. Their first contention is that in a real referendum, voters have to pay the cost if the referendum passes; while in a CV study respondents do not. In other words, a CV study is hypothetical, an issue we have already shown to be an ineffective criticism of the CV method in and of itself. What is important in a CV study is that the scenario it presents be plausible to the respondent. CV surveys are not unique in asking respondents to make choices about the provision of public goods. In an effort to improve the delivery of government services, government agencies, ranging from the local sheriff's office to the Internal Revenue Service, spend millions of dollars each year conducting surveys to find out what the public wants and how much they are willing to pay for it. These surveys have a clear influence on agency actions and hence influence what the public gets and pays. The incentive CV respondents have to reveal their true preferences when answering a survey question is a belief the government is more likely to implement an alternative the more respondents there are who favor it. A binding referendum, an advisory referendum, and an advisory survey all have been shown to have the same properties for truthful preference revelation (Carson, Groves, and Machina, 1999). Our experience with CV surveys is that respondents take them seriously and give answers based on their desire for the good and their budget constraints.

TER contend that a second difference between referenda and CV surveys is that in the latter when respondents are asked whether or not they are willing to vote for a program that will cost them $x, people anchor on the dollar values which are chosen by the survey designer.[6] Anchoring occurs when respondents base their answer to a question on a particular piece of information presented in a survey. The choice of the particular dollar amounts a CV designer presents to the subsamples of respondents in a particular CV study is determined by two considerations: (1) the range of offered prices or dollar amounts must not exceed the range the public finds plausible for the good being valued, and (2) the range and distribution of the dollar amounts (price points) should increase the statistical precision of the estimate. Concern about anchoring seems to arise from the psychology literature which considers "anchoring" to be an undesirable phenomena that leads to biased estimates in studies of human cognition. While undesirable anchoring does occur in some

[6] In CV surveys such as the COS survey that use the dichotomous choice approach, random subsamples of respondents are presented with different dollar values and asked whether they are willing to pay the particular amount presented to them. The resulting set of responses for all subsamples is used to statistically estimate the WTP amount.

situations, the CV survey design used in the COS Study is not one of them; CV researchers *assume* that anchoring occurs; and the estimation procedures they typically use in discrete choice CV surveys are based on this assumption. They randomly assign respondents to different dollar amounts and use statistical techniques that assume a respondents willingness-to-pay decision is influenced by the specific amount she is asked about. This effectively turns the anchoring problem on its head and exploits its structure.

The WTP question in a CV survey asks the consumer to choose either to pay $x dollars to obtain a public good or to keep $x dollars and do without the public good. The information extracted from the response is whether the public good is worth the particular price ($x) assigned to the consumer in the survey design. Respondents need to take the particular $x into account to give a meaningful answer to the WTP question. As such they should be "anchoring" on the amount asked about. Anchoring is only a problem if the respondent is asked about multiple amounts or the statistical technique does not properly condition on the respondent anchoring on the amount asked. In our case, the respondent is asked about only a single amount and the statistical procedures used correctly take account of respondents conditioning their responses on $x.[7] The particular set of dollar amounts presented to respondents in a CV survey does not bias the aggregate WTP estimate as long as one fits the correct parametric distribution to the responses or uses a non-parametric technique.

The third difference claimed by TER is that "the goal of damage assessment is to determine a specific dollar value of foregone services while the goal of a referendum is to determine whether or not some program should be adopted." This statement suggests a misunderstanding of the relevant economic theory. It has long been known that information about public preferences for public goods could be drawn from votes (Black, 1958; Atkinson and Stiglitz, 1980). It has also long been known that if one could observe votes at different effective price levels, estimates of standard Hicksian welfare measures like WTP or minimum willingness to accept compensation (WTA) could be readily inferred (Deacon and Shapiro, 1975; Bergstrom, Rubinfeld, and Shapiro, 1982). The only difference between the voting context and a binary discrete choice CV survey is that the survey designer has some control over the dollar amounts that are presented to the respondents. These dollar amounts are typically chosen to enhance the precision of WTP estimates.

The TER Critique further states that higher precision is needed for damage assessment than for voting; but they offer no reason for this assertion; and it is unclear why this is should be so. While higher levels of precision are almost

[7] The use of subsamples exposed to different price points in this way is similar to dose-response experiments in biology and medicine where different members of a species are subjected to different doses of a chemical. Each subject reacts to one of a range of doses; by combining these reactions the researcher is able to trace out a dose-response curve; and analysis procedures exploit this phenomenon.

always desirable in any decision context, absolute precision is an unobtainable standard in any tort or other economic action such as anti-trust involving damage claims. The degree of precision in the damage estimate has long been a topic of debate among legal scholars (*e.g.*, Landes and Posner, 1987) and is well beyond the scope of these comments. We do note, however, that because the percentage in favor of the program can be observed at the several different dollar amounts chosen by the survey designer, the precision of a CV estimate of WTP will almost always be substantially better than that derived from referendum data because different voters typically face little or no variability in the tax prices in the choices presented in the referendum.[8]

The fourth difference claimed by TER is that "the access to information is controlled by the survey designer in a CV survey; while voters have the opportunity to obtain as much or as little information as they desire in a referendum" [p. 6]. We do not understand why TER consider this a reason against the use of CV surveys. If one takes as the ideal standard a perfectly informed voter, then a respondent in a CV survey is likely to be a much closer approximation to this standard than a voter in an actual referendum. There are several reasons for this. The first is that a good CV survey presents an accurate, balanced, and reasonably complete set of information for the respondent in a manner designed to be easy to understand. Frequently, considerable effort during the development phase of a CV survey like the COS survey goes into determining what most respondents know and don't know about the good to be valued as well as what they think is the important information they need in order to make an informed choice.

Contrast this with the typical election where a voter faces multiple candidates and multiple ballot propositions. For each ballot choice there are competing sides; and each side wants to convey the information most favorable to its position and suppress information that is not favorable. More information, even if truthful, does not lead to better decisions on the part of voters unless it is balanced information. This point has been made recently by several authors in leading journals (*e.g.*, Baron, 1994; Lohmann, 1993; 1994). Also, the amount of time that a respondent spends learning about a particular choice is also typically much longer in a CV study than that spent by the typical voter on a single ballot proposition.

Thus, we conclude that the arguments offered by TER to support their claim that so-called *hypothetical bias* in CV surveys bias their estimates upward are either erroneous or irrelevant to the COS Study.

8 A lack of substantial price variability also adversely influences the precision in estimates based on revealed preference techniques like travel cost analysis, a difficulty compounded when the analyst is forced to infer what choices and prices agents thought they faced when making their decision (Randall, 1994).

Major Flaw 2: Many Respondents Did Not Provide a WTP Response for the Described Scenario

TER seem to regard this purported flaw as the most grievous as they list it first among the "critical" three flaws in their October 1996 transmittal letter. Certainly the failure of respondents to base their WTP amounts on the particular set of contingencies described in a scenario is a problem because such respondents in effect express their value for a different good than the one intended (Mitchell and Carson, 1989; Arrow *et al.*, 1993); and this can result in a biased WTP estimate. Whether this happens or not depends on the complexity of the information, the skill with which the survey designers communicate it, and whether the study design provides data that can be used to test for this type of bias and correct for it if it appears. However, for the estimates to be biased, one needs to have systematic deviations in one direction from the intended scenario. In contrast, random variations in both directions from the intended scenario are unlikely to result in biased estimates but may increase the variability of the estimates (Mitchell and Carson, 1989).

Designers of a CV survey should attempt to minimize respondent misunderstanding and scenario rejection and adopt approaches to ensure that any such misunderstandings do not affect the WTP estimate. Throughout our work in developing and administering the COS survey, we were alert to the possibility of scenario misunderstanding[9] or rejection and sought to minimize it and detect its occurrence in at least four ways. First, we conducted the series of focus groups, individual interviews, and pretests described in Chapter 2 herein and in the COS Report in part to help design a questionnaire which would minimize these effects. Second, we included various diagnostic questions in the survey instrument to help assess the degree of success in this regard. Third, the diagnostic questions included a number of open-ended questions which gave the respondents the chance to ask questions and to explain in their own words why they did or did not vote for the program. Fourth, in our data analysis we conducted tests for possible bias and sensitivity analyses to determine the effect of possible biases on the WTP amount. In all these activities we followed the conservative design principle we have long advocated for CV questionnaire design and data analysis (and which is recommended by the NOAA Panel), which is to consistently make design and analysis decisions which, if they have any effect, will *lower* the WTP estimate. In the COS Report we presented a detailed discussion (repeated herein) of our findings relevant to these issues and included in the appendices all the relevant raw data on which we based our conclusions. TER's assertions that the COS respondents did not provide a WTP response for the described scenario either rest on a weak evidential basis or ignore the sensitivity tests we reported in some detail in our report

9 Surveys are not unique in this regard. Consumer behavior sometimes is based on misunderstandings about the characteristics of the goods they purchase. Indeed, advertising often attempts to exploit these misunderstandings or potential misunderstandings.

which show no upward bias in our WTP estimate from respondents not understanding or accepting the particular scenario elements that concern TER.

TER try to establish several premises to support their contention that COS Study respondents were not valuing the scenario as described. The first is that "many" respondents were valuing protection from a larger spill with larger injuries. Second, some respondents must have been unclear about the injuries due to "unclear" and "ambiguous" descriptions in the COS survey instrument. Third, some respondents thought the program would prevent harm to people. Fourth, some respondents received different information and therefore valued a different scenario. Fifth, the ten-year interval for the injuries described in the COS survey allows respondents to make assumptions about where in the ten year period the described injuries take place. Differing assumptions about the timing of the injury can result in respondents valuing different scenarios. Sixth, respondents were valuing the warm glow of self-approbation from voting to pay taxes. Seventh, respondents were given inadequate information about substitute goods. Eighth, respondents were not given adequate reminders of their budget constraints.

In what follows we discuss each argument and piece of evidence TER adduce to support their claim that our respondents did not value the described scenario. When we talk in what follows about respondents who have not accepted or who have not understood the scenario as intended, we refer to respondents whose understanding and acceptance of the COS scenario differed from it in a substantial way.

1. Assertion that many respondents were thinking of a protection program that would prevent an *Exxon Valdez*-size spill (p. 10).

(1) Evidence from the open-ended questions

An important and distinctive feature of our study is our use of open-ended questions located strategically throughout the questionnaire. Open-ended questions are rarely used in survey research today because they are very expensive to administer (it takes an interviewer much longer to record the essence of a comment than to code the appropriate answer category in a close-ended question) and expensive to code. Coding is the process by which a trained research assistant assigns each segment of a open-ended response to one of a set of relevant answer categories. Our coding procedures are briefly described in section 5.2.1; the coding categories are laid out in Appendix D. We adopted this laborious and costly strategy because the responses of respondents to open-ended questions provide one way to assess the degree to which respondents accepted the premises of the scenario. We not only report the results of the coding process in tables embedded in the text but in Appendix E we also present the raw data on which these findings are based so reviewers might have the opportunity to assess these data themselves. Appendix E includes every comment made by every respondent at any point in the questionnaire.

We would naturally expect TER to draw on this rich body of data in making their assessment of the COS survey instrument. They only draw on it twice. The instance we will consider here is TER's claim that "*many* respondents were thinking of a protection program that would prevent an *Exxon Valdez*-style spill" (italics added). If true, this would be a problem because the *Exxon Valdez* Oil Spill [EVOS] in Prince William Sound, Alaska was the largest spill ever to occur in U.S. waters and the natural resource injuries it caused are many times the size of the injuries described in the COS Central Coast spill scenario. However, it received such a large amount of press and TV coverage in 1989 and 1990 that we would expect some respondents to refer to its effects on wildlife as examples of what can happen when an oil spill occurs. Indeed, the EVOS has assumed the status of the archetype of American oil spills; and respondents could hardly be expected to talk about oil spills without thinking of and mentioning the EVOS. Thus a mere mention of the EVOS does not by itself indicate that the respondent was valuing a EVOS-type spill.

In designing our survey instrument we initially told respondents's explicitly that the prevention program would not prevent spills as large as the EVOS. As noted also in the COS Report, we worried that some of the reasons given for voting for the program in the first pretest suggested that some respondents believed that they were valuing the prevention of spills larger than the program would prevent (see section 2.4.4.2). We concluded on the basis of interviewer comments that explicitly mentioning the EVOS in the way we did must have brought it more powerfully to the respondents minds than would otherwise have been the case. Furthermore, we were concerned that it did this without convincing them that this large a spill is irrelevant to their valuation of the prevention program. We then decided to omit mention of this spill and focus our design efforts on conveying the nature of the spills in the our injury scenario without the EVOS comparison. The reasons for voting given in subsequent pretests convinced us that not mentioning the EVOS minimized possible confusion by respondents that the COS prevention program would prevent spills of that size.

What evidence do TER present in support of their contention that *many* COS respondents were thinking they were preventing another spill of the size of the EVOS? Surprisingly little for such a strong claim: the text of eight responses "selected from responses to the vote-motivation question (Question B-2)" made by those who voted for the COS spill prevention program (p. 10). Each of these statements was made in response to the question which read as follows: "People have <u>different</u> reasons for voting for the Central Coast prevention program. What would the program <u>do</u> that made <u>you</u> willing to pay for it (underlining in the original)?" Interviewers were instructed to use the following probe if the answer did not seem specific enough: "Was there something <u>specific</u> that the program would <u>do</u> that made <u>you</u> willing to pay for it?"

Since the COS Report contained the answers of approximately 550 respondents to Question B-2 [see Appendix E herein], we presume the eight responses TER presented as supporting their argument are the ones that TER believe most clearly make their point. Examination, however, shows that *none* of the eight clearly supports TER's claim that the respondent thought the COS plan would prevent a spill of the same size as the Alaska spill. All eight responses are consistent with another interpretation, that the Alaska oil spill was mentioned because that highly publicized event had *sensitized* the respondent to the possibility of tanker accidents and the *types* of harm the resulting oil spill can cause. For example, one of the eight respondents (Respondent 11012) says:

> I have a big aversion to oil spills after seeing what happened to the oil spill in that sound. [Probe] Anything to protect the shoreline would be worth the $65.

Nothing in this respondent's response suggests that the respondent thought the California spill described in the COS scenario would be as big as the Alaska spill.

Other responses quoted by TER are fully consistent with the sensitization interpretation. Each of the following refer to spill characteristics such as the slow clean up response, slow recovery, harm to the local environment, harm to birds and wildlife:

> (Respondent 11244) Environmental reasons mainly the recollection of *Exxon Valdez* and the slow response for clean up.

> (Respondent 11413) Prevention to unnecessary deaths to any living animals. [Probe] True it is only five dollars. [Probe] Prevention part after seeing what the Valdez did, that area changes are very slow and it causes hardship to surrounding environment.

> (Respondent 11771) Give me peace of mind to protection (sic) of our environment and our wildlife. It is very important to me. I'm somewhat idealistic about this. I keep thinking about the oil spill in Alaska. Oil spills are ugly. Birds and wildlife, all plants have just as much right to exist as we do. $220 is a lot, somewhat borderline for me now. The cost to preventive (sic) is less than the cost of an emergency spill. Prevention is good.

In all but one of the remaining open-ended responses presented by TER, the respondent mentions non-size aspects of the Alaska spill as an example of types of environmental damage from oil spills or negligence by oil tanker captains. Even the remaining one does not clearly support TER's interpretation:

> (Respondent 10981) It's like what happened in Alaska We don't want it to happen here. [Probe] The Valdez thing – it causes a lot of damage. [Probe]

In summary, the claim that *many* respondents thought they were paying to prevent a spill of the size of the EVOS is unjustified. The only evidence presented by TER to back up their assertion is the text of eight of the approximately 550 answers to Question B-2. An examination of the eight responses show that most convey a meaning other than that TER claim; and none unequivocally support the TER contention that "many" respondents were valuing the prevention of a larger oil spill.

(2) *Additional evidence from Questions C-1 and C-3*

Throughout the development of the COS instrument, we devoted a great deal of effort to developing a program that as many respondents as possible would perceive as effective in preventing the specific set of injuries. Given the *ex ante* nature of the prevention program and the probabilistic nature of oil spills, some divergence between the scenario and respondent beliefs is inevitable. Two claims in particular were difficult to convey with certainty: the absence of effects on mammals and on fish, and, especially, the length of the recovery periods (5 years for small animals and saltwater plants and 10 years for the populations of affected species of birds) which respondents tended to regard as short. Similarly, a number of respondents found it hard to believe the program described in the scenario will be *completely* effective given their experience with other State programs. To monitor the degree to which respondent beliefs matched those the scenario attempted to communicate, we designed the set of diagnostic questions in Section C of the research instrument which are asked immediately after the WTP questions in Section B. The answers to the diagnostic questions also provide the necessary information to test whether any divergence between the assumptions presented in the scenario and those actually held by the respondent biases the WTP amount.

TER argue (p. 11) that the responses to C-1 and C-3 are additional evidence "that respondents did not understand or accept the commodity being valued." Question C-1 read as follows:

> Please think back to a few moments ago when I asked you whether you would vote for or against the program. At that time, did you think the harm from oil spills in the Central Coast over the next ten years would be about the *same* as that shown here, a lot *more* or a lot *less*.

Of those answering this question, 34.5 percent said the *same*; 15.7 percent said *a lot less*; 34.8 said *a lot more*; and the remaining 15 percent said they were not sure or gave responses suggesting they were not sure.

TER also cite the responses to another debriefing question (C-3) which asks how effective respondents thought the prevention program would be in preventing harm from Central Coast spills. Here 6.0 percent said *completely effective*; 44.7 percent said *mostly effective*; 38.8 percent said *somewhat effective*; 5.5 percent said *not too effective*; and 5.0 percent said either *not effective at all* or *not sure*.

How do the divergent views about effects size and certainty of prevention affect respondent's willingness to pay for the program? In the COS Report we pointed out that those who believed the harm would be *a lot more* were significantly more likely to vote for the prevention program and those who said *a lot less* were significantly less likely to vote for the program. Those who did not think the program would be *completely* or *mostly* effective were less likely to vote for the program. These are precisely the effects one would expect if respondents found the payment aspect of the choice scenario credible. Furthermore, some respondents may have taken these questions as an opportunity to air their preconceptions even if they had set them aside previously in the face of the information presented in the questionnaire. Thus, these measures may overstate the extent to which respondents actually valued a different good.

To examine whether this level and direction of scenario rejection biases the median WTP amount ($60.56) in any particular direction, we performed the sensitivity checks we reported in the COS Report and in Chapter 6 herein. Our findings are as follows.

Amount of harm. When we estimate a model for assumptions about the amount of harm that sets the value of two dummy variables – MOREHARM and LESSHAR – to zero, the effect on the estimate of median household willingness to pay is only a few cents (section 6.7). This suggests that the upward bias in the WTP amounts contributed by those who thought there would be more harm almost exactly offsets the downward bias caused by those who thought the harm would be less than described. *Thus there is effectively no net bias from this source on our aggregate WTP estimate.*

Effectiveness of prevention program. The construction of two dummy variables for the responses about the perceived effectiveness of the program and setting them to zero models a situation where everyone believes the plan is effective. When we estimate this model the median WTP *increases* to $119.42. This means that if all the respondents believed the plan would be *completely effective* or *mostly effective* our median WTP estimate would be much higher, $119.42 instead of $60.56. *The skepticism some respondents have about the plan's effectiveness therefore exerts a strong downward bias on the aggregate WTP estimate.*

TER's use of the data from our diagnostic debriefing questions in criticizing the validity of our WTP estimate for preventing the oil spill in the Central Region of California is selective. They nowhere acknowledge, much less criticize, the check for bias from scenario rejection we discussed in the COS Report. This analysis, briefly summarized above, clearly shows that there is no net upward bias in the median WTP amount caused by respondents failing to accept the scenario features highlighted by TER.

2. Assertion that our discussion of the oil spill's impact on wildlife was unclear and could have led to some confusion or misunderstanding by respondents

TER criticize the language we use in the survey instrument to describe some of the consequences of the oil spill on wildlife as "unclear" and "ambiguous". Among the examples offered from the COS survey are the statement that marine mammals "are not **usually** affected by the oil because they **generally** leave the area when a spill occurs (p. 11) and the statement that "scientists **expect** that a total of 12,000 birds will be killed by oil spills." TER put in boldface the terms they thought were ambiguous. According to TER (p. 11), "[m]ost likely as a result of the uncertainty in the information provided, respondents in the pretests expressed some skepticism about the number of future spills and the effects of the spills."

CV designers frequently face a tradeoff between using words that are more technically precise and words that respondents find understandable and credible. If the choice whether to pay for the environmental good in a CV survey is not perceived as realistic, respondents may not take at face value all the scenario features, such as spill size, number of birds killed, etc., when they make that choice. In our early pretests, a number of respondents were skeptical about the certainty of some of our assertions about the effects of the spill. Their concerns were not unreasonable: how can we be *absolutely* certain that no mammals will be affected by a Central Coast spill? Contrary to what TER suggest, we found that it was the insistence on certainty in these instances that confused the respondents. The terms which TER label as "ambiguous" appropriately convey the sort of uncertainty in the effects of a spill which those pretest respondents found missing.

3. Criticism that some respondents voted for the program to prevent possible physical harm to themselves or others

According to TER, 8% of the COS survey respondents voted "for the program to 'prevent possible physical harm to respondent or others' (pp. 11–12)." TER's claim is not correct. The 8% TER refer to is from table 5.3; and the percentages in that table are based only on respondents in the sample who voted for the program at B-1 (44 of 552 respondents or 8%), not on the whole sample (44 of 1085 respondents or 4%) as TER seem to think.[10] Second, most of those eight percent (44 respondents) of the votes for the program also gave additional responses in one or more other categories; only one percent of the votes for the program (6 respondents) mentioned health without also mentioning some

[10] Because many respondents gave multiple reasons, the percentages in Table 5.3 add up to 149 percent.

other reason.[11] These 6 respondents are 0.6 percent of the total sample, a far cry from TER's claim of 8%.

In addition, the B-2 debriefing question is followed by a reconsideration question (B-3) which asks respondents how they would vote if there were no human health effects. The question was phrased to encourage even respondents who did not accept that there would be no human health effects to put aside that belief and vote as if there were no human health effects. Twenty-one of the 552 respondents who voted for the program at B-1 were recoded as votes not-for. Six of the 44 respondents who indicated human health was a factor in their vote at B-2 were recoded to votes not-for, suggesting that the other 38 respondents who mentioned human health at B-2 found adequate grounds for voting for the program without human health. The other 18 respondents who changed their votes at B-3 did not mention human health at B-2; changing their votes at B-3 suggests that either their response to B-2 was not complete or that they used the B-3 reconsideration question as an opportunity to change their *votes for* other reasons.

4. Different respondents received different information

TER make the following criticism:

> Different respondents were also given different information regarding the impact of oil spills on mammals. For example, respondents were only told that otters would be unaffected if the respondents specifically asked about otters. [p. 12]

Contrary to the misleading implication of "for example," this is the *only* occasion on which we supplied additional information about mammals. Second, the statement identified by TER as providing different information does not provide significantly different information. In the COS survey *every* respondent was informed that "Marine mammals – such as whales, seals and dolphins – are not usually affected by the oil because they generally leave the area when a spill occurs." This statement refers to "Whales, seals and dolphins" as *examples* rather than as the entirety of marine mammals; and any respondents who thought about sea otters would have most likely lumped them in with the other marine mammals. Just in case any respondent was still uncertain about sea otters and specifically asked if they would be affected in an oil spill, which our pretesting indicated that a few respondents might do, we provided the interviewer with an additional script which read as follows:

> Like other marine mammals, sea otters usually leave the area where a spill occurs. They have not usually been affected by past Central Coast spills.

[11] Five percent (28 of the 552 respondents) mentioned health as one of two reasons; and one and a half percent (8 respondents) mentioned health as one of three reasons.

This information is fully consistent with the information provided to all respondents in the first statement about marine mammals. TER do not offer any evidence that any respondents perceived this information as describing a different good. The very small number of persons who asked about otters (N = 6) suggest that it wasn't much of an issue. TER's mere assertion that it created a difference therefore is not compelling.

5. Failure to specify the timing of injuries

The COS survey asks respondents to value the total amount of harm to wildlife that is predicted to occur in the next ten years during which all single-hulled tankers and barges can be replaced by double-hulled vessels. TER (p. 12) criticize our scenario for not specifying exactly when the injuries would occur during the ten year period. According to them "respondents values ... may be different if the injuries occur in the next year as opposed to ten years from now." TER's comment on this point is correct; however, TER seem to misinterpret the import of this point. The distribution of spill injuries over the ten year period is likely to matter to respondents; but this fact leads to the conclusion that our estimate will be on the low side if it is applied in a damage assessment context since the relevant case in that situation is that all of the spill injuries occur at a particular time whenever the spill occurs. Respondents who believe that some or all of the spill injuries would take place later in the ten year period would be expected to decrease their WTP relative to the case where all injuries occur at the beginning of the period.

We believe that following TER's recommendation that a specific declaration of the time of the spills and injuries would adversely affect the scenario's credibility. That is because it is extremely unlikely that respondents would consider a scenario that predicted the *precise* year a spill will occur and *exactly* what damages it would cause as credible. In contrast, our pretest research showed that describing spill damages over the ten year period it will take to implement the double-hull requirement was widely accepted by our respondents. Second, our leaving respondents to make their own assumptions about when the injury will occur does not harm the validity of the WTP estimates. Individuals value goods with an element of uncertainty all the time.

6. Assertion that Many Respondents gave "Warm Glow" Responses.

TER resurrect the tired and much disputed "warm glow" claim in their review of the COS Study. According to the "warm glow" critique of contingent valuation studies, CV respondents do not pay attention to the specific environmental good or bad in deciding what they are willing-to-pay; instead they express a dollar value for a generalized emotional value they hold for the environment. In their description of the "warm glow" effect, TER overstate the seriousness of the problem by selectively quoting from the NOAA Panel. In addition, the evidence they present to support their contention that the COS Study is seriously affected by "warm glow" is very weak. Finally, they ignore the statistical tests we conducted to test for "warm glow" which clearly show that the median WTP amount expressed by respondents meets the appropriate statistical test.

Kahneman and Knetsch introduced the concept of "warm glow"into the environmental economics literature in a paper (1992) that criticizes the CV method. Based on their skepticism that respondents in surveys are able to process the information conveyed in a CV scenario with sufficient accuracy to make a reasoned judgment about how much a carefully specified environmental improvement is worth to them in dollars, they argue that what CV respondents really express a value for is a generalized "good thing" they associate with the offered good. For example, instead of valuing a specified improvement in water quality in a specific river, respondents value the good thing of "helping the environment." To describe the motivation behind these generalized values Kahneman and Knetsch borrow the concept of "warm glow" from Andreoni who used it in a different context. According to Kahneman and Knetsch, willingness to pay values motivated by the expectation of receiving a warm glow from contributing to a generalized good such as helping the environment do not have economic meaning. One way to test whether people in a CV survey are motivated by economic rationality and value the offered good or are instead motivated by warm glow and value a generalized good is to vary the amount of the good that is offered to separate samples of respondents. The outcome consistent with economic rationality is that CV respondents who are offered a greater amount of a good will be willing to pay a greater amount of money for it. Those motivated by a warm glow will be expected to be insensitive to the scope of the offered good. Kahneman and Knetsch conducted several scope experiments where subsamples of respondents were asked to value greater or lesser amounts of the same public good. In each case the WTP amounts for those who received more of the good did not differ in size from those given by the subsample who received less, confirming their warm glow prediction.

Their claim that what CV surveys value is this type of "warm glow" has recently been challenged on several fronts. First, Chilton and Hutchinson (1999) assert that it misinterprets Andreoni's use of this concept. Andreoni (1989) originated the concept to explain why some people give money to charities that, say, help the poor, when they could free ride on government provision of the same outcome. Andreoni defines warm glow as the premium people receive when they make *voluntary* contributions, such as donating money to fight AIDS; and Andreoni treats it as having real economic value.[12]

12 According to Chilton and Hutchinson (1999), Andreoni disavows Kahneman's interpretation and application of his work.

Second, the motivation to give warm glow responses for government provided goods is low. Champ *et al.* (1997) compare how much a sample of people said they would contribute to a specific charitable organization to improve the area near the Grand Canyon with what a comparable sample of people actually contributed. For a voluntary contribution of this type respondents have an incentive to say they will contribute if they want the good provided: this choice will increase the probability a fund raising effort will be undertaken. When time comes to decide whether or not to actually pay for the good, they will have the option of free riding on the contributions of others. Champ *et al.* found that for survey-estimated donations to exceed actual contributions is consistent with theory. Champ *et al.* argue that their approach of eliciting both a WTP amount in a survey and in actual contributions provides an upper and lower bound on true WTP for the good. Chilton and Hutchinson (1999) note that Champ *et al.* implicitly assume that findings based on the provision of a good by private charitable agencies are applicable to government provision. While respondents may be motivated in part by the warm glow to voluntarily contribute to a private charity and therefore, by Andreoni's reasoning, pay a warm glow motivated premium for the privilege, individuals choosing to pay taxes for the good do not receive a warm glow unless they get utility for paying increased taxes and getting nothing in return. Therefore, Chilton and Hutchinson argue that the lower and upper bounds of Champ *et al.* may overestimate the value of *government* providing the improvement. The reason for this is that there may be a difference between the true WTP for a good depending on whether it is provided by a charitable organization or the government. This difference is the sense in which Andreoni uses the concept.

Although Kahneman and Knetsch (1992) cite Andreoni for support, their warm glow is largely an explanation for their assertion that CV surveys are insensitive to scope. They argue that moral satisfaction/warm glow should be thrown out and that it could not be an economic value. As Harrison (1992) clearly points out in his comment on the Kahneman and Knetsch paper, the motive that lies behind an agent's WTP is simply irrelevant from an economic perspective to a good's value.[13]

Warm Glow and the NOAA Panel

TER quote the NOAA Panel as "arguing that embedding [of which the presence of "warm glow" is an example] is perhaps the most important internal argument against the reliability of the CV approach" (p. 13). They do not, however,

[13] While this is true at the individual level, there may be technical problems with the aggregation of WTP values across individual amounts if certain types of altruism or envy are present. As McConnell (1997) points out, such aggregation problems have nothing to do with CV *per se* and are unlikely to be involved in studies such as ours where agents are informed that other agents will also be taxed and altruism is likely to be directed at the animals or ecosystem involved rather than other agents.

mention that the NOAA Panel rejected the view promoted by critics of the continent valuation method that "warm glow" responses are inevitable. Instead, the NOAA Panel posed the question "... how should a CV instrument be framed to elicit an answer that responds to the precise scenario and not to a generalized 'warm glow' effect?" and answered it by recommending a "high standard of richness in context to achieve a realistic background."

TER's Evidence of Warm Glow

TER base their claim that "many respondents were ... giving warm-glow responses rather than their true vote for the prevention program" (p. 13) on the basis of a "careful review" of the B-2 open ended responses. In order to draw any conclusion about "most respondents" such a review would require at the very least a systematic coding of the full set of responses according to specific and defensible criteria capable of differentiating a "warm glow" response from responses not evincing a warm glow. TER present no criteria and do not describe what their "careful review" consisted of. Instead, they present a list of eight "examples of some embedding and warm-glow responses ..." from the entire set of approximately 450 open-ended responses to question B-2. An examination of these examples does not support their interpretation.

When we examine the text of the eight responses, it is important to keep in mind the context in which B-2 is asked and the manner in which respondents are likely to respond. Respondents have just been given a lot of information about the injuries the program will prevent and then they have voted in favor of the program. Now they are asked question B-2:

B-2 People have *different* reasons for voting for the Central Coast prevention program. What would the program *do* that made *you* willing to pay for it?

The respondent will assume the interviewer is familiar with the effects described in the scenario and will take these for granted in what he or she says the program will do. Thus a general statement such as "the program will help the environment" is probably intended to be understood in the context of the information about the specific things the program would do that was shared with the interviewer (Section 5.2.3.1). In order to minimize this conversational parsimony effect, the interviewers were instructed to probe further about any response expressed in general terms to see if they had any of the specific effects in mind. The wording of the probe was "Was there something specific that the program would do that made you willing to pay for it?"[14]

We discuss conversational parsimony and other conversational conventions

[14] TER are mistaken (p. 14, note 14) when they say the COS Report does not contain the exact text of the probe we used here. Its complete text may be found on page 3–18 of the COS Report and on page 15 of Appendix A of the COS Report which contains the COS survey instrument.

(Grice, 1975) in some detail in Chapter 5, section 5.2.3.1. In order to test the applicability of the conversational parsimony theory to the answers to the B-2 question, we conducted the experiment described in that section. In this experiment, 94 respondents were randomly assigned to a treatment condition where Question B-2 was asked by someone who was *not* present during the previous part of the interview, and 84 to a control group which replicated the condition where the interviewer administered the entire interview. The results showed that those in the treatment condition used more words to explain their votes *for* to someone who had not taken part in the interview compared with the number of words used by respondents in the control group. This confirms the prediction from conversational parsimony that respondents explaining why they voted for the program to someone other then the interviewer will have more to say. While there is much about conversational parsimony this experiment does not explain, it does demonstrate that respondents are more verbally forthcoming when they presume the person with whom they are conversing does not share their knowledge of the topic.

TER call the results of our experiment "questionable" on three grounds. First, they claim (p. 16) that the difference in the number of words used in the two conditions could have been "the result of a small sample size." TER do not further elaborate. Since we reported the difference was statistically significant, and since the statistical test takes the size of the samples into account, this comment seems to be based on a lack of understanding of statistical testing.[15]

Second, they claim that our analysis counts words rather than examines their meaning. We did not necessarily expect *more* reasons but rather a richer explanation that might manifest in several ways and even vary among respondents. The number of words is a reasonable operationalization of this variable.

Third, TER complain that we "do not explain how [we] take into account [the effects of conversational parsimony] when coding the open-ended responses [to Question B-2]." This is a puzzling criticism because the coding process we describe in our report (section 5.2.1 herein)[16] does *not* take conversational parsimony effects as such into *account* at all. Only after the responses are coded do we draw on conversational parsimony for insight about why the respondents explained their *vote for* the program in the way they did.

We now turn to the eight responses to question B-2 that TER regard as

[15] As we report on pp. 5–9 of the COS Report, the difference in the average number of words used by the treatment condition (61 words) compared with the control condition (50 words) is statistically significant at the .02 level for using a one-sided t-test. TER assert that our result may be an artifact of a small sample size; in fact it is *more* rather than less difficult to achieve statistical significance when sample sizes are small.

[16] The categories into which the coders placed the various ideas expressed by the respondents were based on the ideas expressed in the *open-ended responses themselves*. The coders who constructed the coding categories and performed the coding were not apprised of either the conversational parsimony theory or of the expected outcome.

showing that "most respondents" in the COS Study "were not valuing the specific services provided but rather giving money for some broader environmental commodity or making a general charitable contribution."

Two[17] of the eight responses cannot be considered "warm glow" responses because they refer to specific services:

(Respondent 10766) Save some birds and wildlife. I'm a bird and wildlife lover. {Probe} That's about it.

Comment: This is exactly what the plan would do; save some birds and wildlife. The revelation that this respondent is a "bird and wildlife lover" has nothing to do with "warm glow." It simply indicates that the respondent has a preference for these creatures, a preference that, according to economic theory, should incline one toward willingness to pay.

(Respondent 10523) I like nature and I like the environment clean. The habitat of the animals should remain. The ducks have no place to return to. They go midwest now. They don't want to go to a polluted area. {probe} {For program specifics}.

Comment: The first statement – I like nature and I like the environment clean. – is fully consistent with the parsimony theory. The respondent assumes the interviewer knows what the program would do and begins by referring to why the respondent valued the program. Then the respondent turns to the service the habitat that would be harmed by a spill provides the birds.

Four of the remaining six responses to question B-2 may be said to provide evidence that the respondents interpreted the request as something more than a request to repeat what was asked; the four respondents begin their responses by explaining their motivation in general terms (knowing that the interviewer is fully aware of the specific services the program would provide). Then, in response to the probe – "Was there something *specific* that the program would *do* that made *you* willing to pay for it?" – the respondent mentions a specific service.

(Respondent 10748) The earth is no different from how we are. The way we view the earth's body and our bodies are the same. {probe} It would show an attitude of concern for the environment, and that is a plus. {probe} To not see more birds covered with oil would do it.

Comment: The specific service is not seeing birds covered with oil.

(Respondent 10289) It would make the oil companies responsibly function. {probe} It is a stepping stone in saving our environment. {probe} Save the birds and wildlife.

Comment: "Save the birds and wildlife" from harm (implied) is the service.

(Respondent 10500) *The $5 is a relative [sic] insignificant tax increase to do anything.*

[17] TER occasionally present only parts of open-ended responses from the COS Report. Here and elsewhere we present the text of the entire open-ended response first presented in Appendix D of the COS Report.

{*probe*} I treasure the natural environment and see it under assault. {probe} Anything we can do to pressure the natural environment is worth the expenditure of public movies (sic). {*probe*} *This looks like a quick response using good technology to lessen the impact of the spill.* [Italics added.]

Comment: TER did not include the text in italics when they presented this response. The answer following the second probe clearly describes a specific service the program would provide.

(Respondent 10608) Anything that helps fight pollution of the environment, even though it might be minimal it's worth $5. {probe} It would prevent oil spills from [harming] wildlife. {probe}

Comment: The reply to the probe puts the opening statement in perspective. When reminded that the question (B-2) asks "Was there something *specific* the program would *do* ...", the respondent unambiguously states a specific service.

Only two of TER's eight examples remain. Unlike the other six examples in which the respondents mention specific services, these two responses never move beyond why they have value to what specific services they have value for. Since conversational conventions dictate you don't repeat shared context, the respondents may have interpreted the request as a request for more information about their motivations for their values; and unlike the earlier examples, repeated probing did not refocus the respondents on the specifics of the scenario.

(Respondent 10552) Have the beaches cleaner, the water cleaner, the atmosphere would be better. Our children would have something nice around them for the future. {probe} All the species there wouldn't go extinct and it would be a cleaner environment for the birds and for us. {probe} That's about it.

Comment: This response is consistent with the interpretation that the respondent is giving money for some broader environmental commodity in the sense that the respondent mentions services the program does not claim to provide such as "better atmosphere" and keeping "all the species" affected from going extinct. However, the respondent does mention a "cleaner environment for the birds" which is a service the program would provide.

(Respondent 11463) I feel all Californias (sic) need to share the responsibility of keeping the earth safe. *Oil spills break the ecological chain. It [sic] affects animals and ultimately humans.* {*probe*} {*probe*}

Comment: If the first sentence of this response was the entire text, it would reasonably fit a "warm glow" interpretation. However, the remainder of this respondent's comment, which TER edited out of the response, focuses on the harm caused by oil spills. Although this harm is expressed in general terms rather than specifically related to the program's benefits, it is possible to infer the respondent has in mind the harm from the oil spills described in the COS scenario.

We discuss these open-ended responses to question B-2 in such detail because they constitute the sole evidence TER present for its "warm glow" claim. We noted that TER had a very large pool of responses to Question B-2 available to them in our report so the eight they chose presumably were those that most clearly made their point. To summarize, of the eight examples of "warm glow" responses TER chose as evidence of "warm glow," six clearly refer to relevant specific services the respondent expected to receive as a result of paying for the oil spill prevention program. They do not support the "warm glow" hypothesis. The fact that some of these responses begin with a general rationale for why the respondent values environmental improvements in general before mentioning the specific services the COS protection program would provide is explained by conversational parsimony. Conversational conventions dictate that one doesn't restate shared context unless necessary so a question seeking such restatement may not be immediately understood as such. The remaining two responses are less specific about environmental injuries being valued; so it is possible, but by no means certain, that these respondents were valuing a service other than those specified in the survey.

7. Adequacy of Substitute Reminders

In section 2.4 of their report, TER criticize the COS Study for inadequately reminding respondents of possible substitutes for preventing oil spill related injuries along California's central coast. In fact the COS survey carefully reminded respondents about possible substitutes. The COS survey begins by calling respondents' attention to other public goods. At the beginning of the interview respondents are asked two sequences of questions A1(a)-A1(f) and A2(a)-A2(f) that elicit respondent views about the desirability of a range of public goods from improving education in California elementary and secondary schools to reducing air pollution in California cities, to providing shelters for the homeless, to building new prisons. The COS survey instrument informs the respondent that these are only a few of the things that the government spends money on and that proposals for new projects are often made. Later in the interview, in the important location immediately before the WTP question, respondents were reminded that they "might prefer to spend the money [that would be used to pay for the Central Coast oil spill prevention program] to solve <u>other</u> social or environmental problems <u>instead</u>" [underlining in original interview text]. At this same point in the interview, private good substitutes are dealt with by reminding the respondent that the program might cost more than their household wants to spend for the proposed program.

TER raise a number of specific criticisms at this point in their critique, including the need to remind respondents of the abundance levels of other close substitute resources, the need to mention animals unaffected by the spill, and the need to place the number of birds killed and number of shoreline miles impacted in relative context. However, an examination of the COS survey shows that either we explicitly addressed these issues or we adopted alternative strategies which are as satisfactory or superior to the approach suggested by TER.

With respect to bird populations, COS show card C clearly displays the California populations of the potentially impacted birds (130,000 western gulls, 290,000 Pacific loons, 250,000 rhinoceros auklets, 525,000 common murres and 140,000 Brandt's cormorants). The survey text further emphasizes the population numbers for particular species. The first of the "Reasons [for voting] Against" the program offered on Show card G is that "[t]he number of birds and other wildlife the program would protect is small in comparison to their total numbers, and none are endangered." (Underlined in the original to cue the interviewer to emphasize those words.) This statement is easy to understand and, in this case, is an effective way of providing information on the percentage of birds that are killed.

TER also argue that "the survey does not emphasize enough that the prevention plan will protect only ten miles of the Central California Coast." (P. 21) In fact this point is carefully emphasized in the COS instrument. In the summary of the expected harm from oil spills off the Central Coast on Card D and the associated text in the COS survey, for example, the text states that in addition to injuries to birds **"MANY** [small animals and plants would be killed] along about **10 MILES** of shore" (the capitalization and bold type is used on the show card). Later, the respondent is shown Card G which summarizes the reasons that a respondent might want to vote for the program. Here, again, the 10 mile figure appears in bolded caps; and the survey text at this point reemphasizes the 10 miles. One possibility for TER's confusion on this issue is their assertion that "the map may have influenced respondents to mistakenly think that the entire central coast area or the entire coast of California would be protected under the program." (p. 21). Here TER refer to the map shown to the respondent on Card D. This map of the state is needed to prevent any confusion between a prevention plan for the Central Coast and one for all of California and to show how the proposed response centers are strategically located to cover the entire Central Coast. It also shows respondents that there are repeated sections of similar habitat along the coast.

TER assert (p. 21) that we should have told respondents that only 2% of the coastline was being protected because the Central Coast is 500 miles long and the spill we describe to respondents that would take place without the plan would only harm ten of the 500 miles of coast. This confuses the specific coastal injury of the prevented spill (10 miles) with the amount of coast that our prevention plan would protect (500 miles). Since the anticipated spill could occur *anywhere* along the 500 mile Central Coast, we are correct in stating that our prevention plan will protect the entire Central Coast from experiencing such a spill. In other words, the COS respondents were valuing preventing the deaths of the described number of birds off shore plus the deaths of "many small animals and saltwater plants ... along a total of about ten miles of shoreline" anywhere on the Central Coast.

8. Adequacy of Budget Constraint Reminder

TER's criticism of the COS budget constraint reminder is that it needed to be more explicit and more focused on the respondents' income constraints. But TER offer no evidence that our budget constraint reminder was ineffective. The COS survey is quite clear that the money would be collected from households via taxation, which is a payment mechanism known for its effectiveness and desirable theoretical properties. We believe the wording we use on show card G – "The program might cost more than your household wants to spend for this" – is more effective than TER's suggested rewording which substitutes "can afford" for the last five words: our wording helps to legitimate *multiple* reasons for not favoring the program, including that there are private goods that are substitutes.

TER fail to note that we offer the respondent two opportunities to switch from favoring to not favoring the program, both of which emphasize the cost of the program. TER also fail to note in this section of their Critique that we took the very conservative step of treating the votes for the program of every respondent who is not currently paying state income taxes as votes not in favor of the program even if they said they were willing to pay the asked amount. We did this because of a concern that our payment vehicle of "higher taxes" might not have the appropriate incentive properties for this group of respondents. TER further fail to note that the construct validity equation in Table 6.7 displays a very significant income effect, an unlikely result if respondents generally ignored budget constraints. Also, our point estimate for the possible percentage of *yea-sayers* is effectively zero, again an unlikely result if respondents generally ignored budget constraints.

Major Flaw 3: The Failure to Test for Sensitivity to the Scope of the Scenario

TER argue that the COS Study should have included a test of sensitivity to the scope of the injuries using independent samples. TER reject our within-sample test of scope and paradoxically claim that the divergence of beliefs of respondents that form the basis of our within-sample test of sensitivity to scope are, instead, evidence of a rejection of the scenario only. The TER Critique compares the COS Study results with the results of two other CV studies, arguing that the estimates of the three studies are evidence of insensitivity to scope.

Regarding the desirability of the COS Study including a between-sample test of scope, TER interpret the NOAA Panel Report to say that any CV study that does not carry out a formal out-of-sample scope test is inherently flawed. This is not the case. While the Panel does list "[i]nadquate responsiveness to the scope of the environmental insult" as requiring a mandatory showing, it does not provide any description of a scope test. At most the panel states that "some form of internal consistency is the least we would need to feel some confidence that the verbal answers corresponded to some reality". Certainly, the Panel would agree that the optimal test of scope would be different scenarios administered to different samples. At that time few CV studies had demonstrated scope in this way and some critics of contingent valuation argued that the method was so flawed that few if any CV studies would be able to do so. Since then, however, a large number of between-sample scope tests have been conducted and analyzed; and the vast majority of these tests reject the null hypothesis of insensitivity to scope (Carson 1997a).[18]

TER nevertheless conclude that only a between-sample scope test will suffice to demonstrate the validity of the COS Study and that without it the COS Study is flawed (p. 17). In response, we should point out it costs a great deal of money to conduct scope tests with separate samples and money spent for an unnecessary separate sample weakens the overall quality of the estimates because it reduces the sample size for the main study. We present an internal scope test in our report (p. 6-30). It shows, at a statistically significant level, that those respondents who thought oil spills along the Central Coast over the next 10 years would cause more harm than that described in the survey should be and were more likely to vote for the program and those who thought that oil spills would cause less harm were more likely to vote *not-for* the program. As noted in the previous section, standard economic theory on information processing suggests that people take their prior information and update it with new information as provided by our CV survey. They will use this combined information set in formulating their response to the WTP question. One would expect then that a respondent's WTP will vary with their posterior expectation of injuries and the likely success of the program. This is indeed the case as shown in Table 6.7 herein and also in the COS report.

The TER Critique adopts conflicting views (p. 17) on internal tests of scope, alternately arguing that internal tests of scope are pointless and then criticizing

[18] Studies failing to demonstrate sensitivity to scope tend to fall into one of two categories. The first category includes studies by researchers sponsored by Exxon and the oil industry. Since that time, several of the claims of scope insensitivity made concerning particular studies (*e.g.*, Diamond *et al.*, 1993; Schkade and Payne, 1993) have not held up under a more considered analysis (*e.g.*, Carson, 1997; Carson and Flores, 1996). Further, disclosures that considerably more studies were sponsored by Exxon than were reported at the 1992 Exxon-sponsored symposium on CV (Hausman, 1993), in spite of assertions made by the organizer of that conference that all studies had been disclosed, has called into question the integrity of these studies. For instance, at the Exxon symposium, Charles River Associates, one of Exxon's subcontractors reported on the results of 400 interviews at five locations; while a subsequent newsletter (Charles River Associates, 1992) inadvertently disclosed it had "conducted a number of experiments, requiring many different questionnaires administered to several thousand respondents at over 30 locations throughout the United States."

The second category of studies failing to demonstrate sensitivity to scope includes CV studies valuing changes in low level mortality risks to humans. This finding should not be surprising as studies based upon observed behavior have consistently found problems with how people react to such changes; and there is a large literature on risk communication that suggests that conveying changes in low level risks is a very difficult task.

the internal test of scope we performed on the grounds that it suffers from what TER call "a sensitization bias." According to this view, a respondent who is asked about one level of a good is sensitized so that when he or she is asked to value another level of the same good there is a pressure towards self-consistency which may not be real (Kahneman and Knetsch 1992). There has been a continuing debate (*e.g.*, Smith, 1992; Harrison, 1992; Carson, 1997a) over this assertion. Be that as it may, the TER Critique fails to recognize that because the internal scope test we conducted is based on information provided by the respondents *after* the only valuation question, it bypasses the source of the alleged insensitivity while exploiting some of the statistical power inherent in being able to control for many of the respondent characteristics. This is not a new approach. In fact it was used in the Silberman, Gerlowski, and Williams (1992) paper cited by the TER Critique to show that respondent WTP amounts were related to the distance they lived from a recreational beach.

In an effort to support their claim that CV studies are insensitive to scope, TER argue that the damage estimates we measure in three of our studies, the COS Study, the EVOS Study (Carson *et al.*, 1992), and the Southern California Bight [SCB] Study (Carson, Hanemann *et al.*, 1994) show the same WTP. TER do not mention the formal between-sample scope test in Carson, Hanemann *et al.* (1994) which showed the smaller injury scenario received substantially lower WTP. TER argue in section 3.3.1 of the their critique that Turnbull estimates of the lower bound on the mean are sensitive to the choice of the dollar amounts used. In section 2.3 of their critique, however, they compare Turnbull lower bound mean estimates based upon different sets of design points rather than focus on parametric estimates of the mean and median WTP since those estimates (although not their confidence intervals) do not depend upon the particular design points used.

Putting aside the inappropriate application of statistics and accepting TER's contention that the two estimates are comparable for the sake of discussion, the TER Critique never makes clear why average WTP for prevention of another EVOS for the nation as a whole should be larger than the average WTP of Californians to prevent a smaller set of injuries along some of the most scenic and most often visited coast line of their home state. One is thus looking at how a simultaneous shift in location (*i.e.*, from out-of-state and not that well known to in-state and very well known) and a shift in injuries (*i.e.*, very large at one time versus small repeated) influence WTP.

Major Flaw 4: Use of a Flawed Analytical Approach for Estimating Average WTP

TER's major focus in this section is to assert that our use of the Turnbull estimator of the lower bound on the sample mean (Turnbull, 1976) for estimating average WTP is a flawed analytical approach. TER's argument on this point makes little sense unless the point TER are trying to make is that our estimate of WTP is likely to be biased substantially downward from its true value. If this is the case, we agree with TER. We do not, however, see this as a flaw in the study. Rather, we see the use of the Turnbull estimator, as part of the adoption of an overall design and estimation strategy that is intended to result in a defensible conservative estimate of the economic damages which occur due to injuries to natural resources. Such a strategy is recommended by the NOAA Panel. The TER Critique also raises two secondary issues in this section. The first is the possibility that we conducted inadequate tests for the presence of *yea-saying*. The second involves our use of a Box-Cox functional form rather than the Turnbull estimator in our construct validity equation (Table 6.7). We discuss both of these issues below.

We cannot determine whether the TER clearly understand the properties of the estimated Turnbull lower-bound on mean WTP. TER's major complaint appears to be that this estimate is sensitive to the design points (*i.e.*, the dollar amounts offered to respondents) chosen by the survey designer. TER assert correctly, for instance, that if the respondents who received the largest design point ($220) are dropped, the estimated Turnbull lower-bound on mean WTP drops 37%, from $84 to $53.[19] TER also assert correctly that the Turnbull lower-bound on mean WTP would likely increase if an additional design point at a higher dollar amount had been used, although there is no evidence from the survey to support the magnitude of the specific increases claimed by the TER Critique.[20] This is a trivial observation because as long as there is one respondent who is willing to pay more than a study's highest design point, there is always some additional design point that would increase the WTP estimate.

It is perhaps useful to step back from the TER Critique and look at how the Turnbull lower bound on mean WTP is calculated in the simplest case. First, assume that all respondents are asked about only one dollar amount, $10. If 60% of the respondents indicate they are willing to pay the $10, the Turnbull lower bound on the sample mean is $6 (.6 × $10). This calculation assumes that anyone who did not say they were willing to pay $10 is willing to pay $0, *i.e.*, nothing. It also assumes that everyone who said they were willing to pay $10 is willing to pay only $10. Both of these assumptions are the most conservative ones that can be made that are consistent with the data.

We do not know if any of the respondents who said they were not willing to pay $10 are willing to pay some amount between $0 and $10, but that is likely. We also do not know if some of the respondents who indicated they

[19] This should not be seen as an argument for dropping the highest design point but rather as a reflection about how the Turnbull estimator works. The properties of the Turnbull estimator are discussed at length below.

[20] As discussed below, this point does not hold for the shift of a design point value rather than the addition of a design point.

were willing to pay $10 are willing to pay more than $10, but that too is likely. If either some of the respondents not willing to pay $10 are willing to pay something greater than zero or some of the respondents willing to pay $10 are willing to pay more than $10, then the true mean WTP will be greater than the Turnbull lower bound on the mean. As survey data does not reveal anything about how much more, the Turnbull lower bound on the sample mean forgoes any assumptions about how much more.

In this simple case, one can also see the impact of changing the $10 design point to a lower amount, say $9. Everyone who was in favor at $10, should also be in favor at $9. Assuming that none of the respondents who were not in favor at $10 are in favor at $9, the Turnbull lower bound on the mean is now $5.40 (.6 × $9). However, the other extreme which is consistent with the original response of 60% in favor at $10, is that 100% are in favor at $9. If this is the case, then the Turnbull lower bound on the mean is $9. Thus, the shift in the design point from $10 to $9 can result in a change in the Turnbull lower bound on the mean from $6 to somewhere in the range of $5.40 to $9. Thus, the implication that TER's attempts to draw in Section 3.3.1 that increasing the dollar amount of the design point *always* increases the Turnbull lower bound mean is incorrect.

Adding an additional design point can only increase the Turnbull lower bound on the mean; but adding an additional design point decreases the precision with which the Turnbull lower bound on the mean is measured (if the sample size is not also increased). As the number of design points increases to be arbitrarily large with very large sample sizes at each of these design points, the Turnbull lower bound on the sample mean approaches the true mean from below. That is, the Turnbull lower bound on the mean is always equal to or lower than the true sample mean that one is trying to measure.

Underlying this result that the Turnbull lower bound on the mean is always equal to or less than the true mean relies on two key assumptions. First, that there are large equivalent samples of respondents at each of the design points used. This is achieved by randomly assigning respondents to design points. The second is that each sample at each design point provides information about what percentage of the population is willing to pay that amount. The Turnbull estimator imposes one key restriction, that the percentage willing to pay cannot increase as the dollar cost increases. (This restriction is discussed at more length below).

The Turnbull estimator uses the percentages at each design point to trace out the percentage willing to pay as a function of the cost in dollars. For example, if the Turnbull's estimated percentage in favor at the first design point, $10, is 60% and its estimated percentage at $25 is 35%, then the Turnbull lower bound on the mean is calculated by assuming that 40% of the sample is assumed to be willing to pay $0 (100% of sample assumed willing to pay $0 and 60% of the sample estimated to be willing to pay $10, leaving 40% who are only willing to pay $0), 25% of the sample is willing to pay at most $10

(60% of sample willing to pay $10 and 35% of sample willing to pay $25, leaving 25% who are only willing to $20), and 35% of the sample are willing to pay $25. Thus, the Turnbull lower bound on the sample mean equals $11.25 [(.4 × $0) + (.25 × .$10) + (.35 × $25)]. There are two key properties of the Turnbull lower bound on the mean that should be noted. First, no respondent is ever assumed to be willing to pay more than the dollar amount of the highest design point used in the survey. Second, the fraction of respondents estimated to be willing to pay the dollar amount at one particular design point but not the dollar amount at the next highest design point are assumed to be willing to pay only the dollar amount of the lower design point. These two properties ensure that the Turnbull lower bound on the mean cannot exceed the latent (sample) mean.

A good choice of design points, which CV researchers use pretests and pilot studies to help determine, can result in a Turnbull lower bound mean that is as close to (albeit still less than or equal) the true mean as possible given a fixed number of design points. There is no choice of design points that can make the Turnbull lower bound mean exceed the true (sample) mean. It is however possible to make the Turnbull lower bound on the mean arbitrarily small by making all of the design points have small dollar amounts. For instance, if there were five design points of $0.01, $0.25, $0.50, $0.75, and $1.00, then the largest possible value of the Turnbull lower bound on mean WTP is $1.00, which would occur if 100% of the sample at each design point were willing to pay the amount asked about. In contrast, it is possible to get a Turnbull lower bound on mean WTP estimate of $0 by making all of the design points so large that no respondents are willing to pay the amounts asked about. As TER point out, the value of the largest design point can have a large influence on the magnitude of the Turnbull lower bound mean estimate. Indeed, a good choice of design points typically results in such a finding. That is because having two design points spaced so that there is little difference in the percent voting for contributes very little to changing the Turnbull lower bound mean estimate. Further, for a fixed number of design points, placing the last design point farther and farther out can often result in a lower, not higher, estimate of the Turnbull lower bound on the mean, as what matters is the percentage willing to pay the highest amount asked.

The TER Critique puts forth one philosophical argument against the use of the Turnbull estimator:

> The Turnbull approach sharply departs from the trend in the CV literature over the past 20 years. During this period CV practitioners have developed more and more sophisticated models to try to find a behavioral structure underlying respondents' CV responses. These researchers sought to construct utility-theoretic models that explain the variation found in these data, and then used those models to estimate WTP measures for policy purposes. These utility-theoretic constructs are important because they require CV data to conform to the same theories applied to other data on economic choice [pp. 25–26].

This argument is just a bit strange in light of the shift away from parametric statistical estimation techniques toward non-parametric estimation techniques in all areas of applied microeconomic research, and, in particular, those dealing with welfare estimates (*e.g.*, Hausman and Newey 1995, Stock 1991).

The reason for this shift is straightforward: the parametric approach makes strong functional form assumptions that can have a very large influence on the resulting estimates; and those functional form assumptions are usually difficult and costly, if not impossible, to verify. Non-parametric estimators do not impose strong functional form assumptions. Their drawbacks are that they are often computationally more difficult to implement and it is often difficult to impose restrictions suggested by economic theory. The Turnbull estimator has neither of these problems; but the Turnbull estimator's weakness is that it does not provide estimates about how the WTP distribution changes between design points or outside the range of the lowest and highest design points.

The TER Critique makes the following statement:

> The COS Report provides little discussion of the relative merits and shortcomings of the Turnbull approach. Given that this approach is new and untested in the CV literature, this omission is remarkable [p. 26].

On this point we believe that TER have failed to adequately review the literature related to this estimator. The COS Report contains many of the relevant citations as well as a fairly comprehensive mathematical treatment of the Turnbull's characteristics (see Appendix E.1 of the COS Report, Appendix F herein). Here we only note the dramatic increase in the use of the Turnbull estimator in the biometrics literature (*e.g.*, Lindsay and Ryan, 1998) as well as substantial recent use of the estimator in the contingent valuation literature (*e.g.*, Carson, Wilks, and Imber (1994), Carson, Hanemann *et al.* (1994), Haab and McConnell (1997; 2002), Boman, Bostedt, and Kristrom (1999), Hanemann and Kaninen (1999)). We further note that first use of the Turnbull estimator in the CV literature is the 1990 Kristrom paper which uses the singled-bounded version of the estimator proposed early in the statistics literature by Ayers *et al.* (1955) and in Carson and Steinberg (1990) who use the general version put forth by Turnbull (1976). As noted earlier, the major advantage of this estimator for use in valuation work is that it only imposes the one restriction of economic theory that the percentage WTP does not rise with increases in price. Further, with random assignment of respondents to design points, there is no need for covariates to describe the unconditional WTP distribution which is all that is needed for many policy purposes.[21]

[21] It is of course, possible to use a Turnbull-type approach with covariates as Heckman and Singer (1984) have done in labor economics if one is willing to impose some structure on the link function between the dependent variable and its predictors.

COS Turnbull Standard Error Calculation

TER assert that "[s]tandard errors give a false sense of precision of the estimates." They continue:

> The COS Report presents standard errors for the Turnbull means that are quite small, but misleading. ... The standard error should reflect two components: measurement error and sampling error. However, the standard error from the Turnbull approach only reflects sampling error [page 28].

It is not clear to whom TER believe that standard errors give this false impression. TER's representations notwithstanding, standard errors, whether those reported in the COS Report or those reported in other disciplines, do not generally purport to reflect anything but sampling error. Stokes and Belin (1998), in the American Statistical Association series on statistical practices, explains the issue this way:

> Why isn't the margin of error adjusted to reflect both sampling and nonsampling uncertainties? The answer is that, unlike sampling error, the extent of nonsampling error cannot usually be assessed from the sample itself, even if the sample is a probability sample.

Thus despite TER's apparent belief otherwise, in substance, TER are not ascribing anything more to the COS Study than that it adheres to standard practice in reporting sampling error. If TER's intent is call into question the wisdom of this practice, they should do so explicitly.

As to the issue of measurement error and other non-sampling error, the COS Report adheres to our past practice of providing a rich source of relevant material for this assessment as exemplified by Carson *et al.* (1992), which the NOAA Panel cited as an example of good practice in providing relevant information in the report of a CV study.

Finally, the standard errors reported in the COS Report are small. A couple of reasons account for the small size of these standard errors. First, the sample size is fairly large. Second, one can get a more precise estimate of the lower bound on the mean than on the estimate of the mean itself because most of the uncertainty around the mean lies on the upper end of the confidence level.

The Potential for Yea-Saying

In their section on yea-saying, TER cast the COS Study as one of several CV studies which exhibit what they call "unexpectedly high numbers" of affirmative votes at the high bid amount. They suggest that this is the result of yea-saying; and they dismiss the test for yea-saying reported in the COS Report. They also argue that the dramatic variation in the Turnbull estimate produced by assuming different percentages of yea-sayers ranging from zero to the number of affirmative votes at the highest bid amount requires "far more justification".

Finally, they argue that the Turnbull lower bound mean is deficient as a statistical estimator because the procedure can be profoundly influenced by large outliers.[22]

TER begin the substance of their argument with a largely implied suggestion that CV studies that use a dichotomous-choice question format generally exhibit an "unexpectedly high" proportion of affirmative votes at the high design points. TER advance yea-saying as one explanation for this purported price-insensitivity and then more-or-less adopts yea-saying as the explanation. Their list of studies with purported yea-saying include several studies in which the authors of this rebuttal participated. Contrary to TER's assertion, most of the studies presented by TER do not seem to have an implausible percentage of affirmative votes at the magnitudes of the design points presented. The two studies on the list which do show a high proportion of affirmative responses at very high magnitude design points are part of the legacy of the Exxon-sponsored attack on CV. In contrast to the other studies TER include in their list, the Exxon-sponsored studies do exhibit what seems an implausibly high percentage of affirmative votes for the magnitude of the design points.[23]

The highest design points presented for most of the studies not sponsored by Exxon are not even in the same ballpark as those in the Exxon-sponsored studies. For example, the percentages of affirmative votes that TER present for the SCB Study and the Selway Wilderness study are 25% and 26% respectively. But the highest design point TER present for the SCB study, a study not sponsored by Exxon, is $215 compared to $2000 for the Selway wilderness study, a study which was sponsored by Exxon. The difference is striking: 25% at $215 is not implausible whereas 26% at $2000 seems quite implausible.[24]

Furthermore, the measure that TER report for the SCB study is not the most appropriate one for addressing the possible frequency of yea-sayers or for this comparison. This is also true of the measure that TER report for the EVOS Study. In fact, if the percentages from the Exxon-sponsored studies do not seem inconsistent with the percentages of the studies with the lower amounts; it is only because of TER's inapt selection of measures. When other measures are reviewed, the inconsistency of results from those studies not sponsored by Exxon with those of the two studies sponsored by Exxon is apparent.

22 Underlying their discussion of yea-sayers and nay-sayers are TER's ill-conceived positions on dichotomous-choice data and the Turnbull estimator. We discuss both these issues elsewhere in this rebuttal.

23 We suspect that the difficulty in these studies occurs because the highest design point asked about would be perceived by many respondents who received it as implausible for a government project of the sort described. In such as case, it may be optimal for respondents to ignore the stated amount in favor of their own assumption about price.

24 We acknowledge that if one assumes that the 25% at $215 presented for the SCB Study did not drop with further increases in the design amounts, these two results would not be inconsistent.

The first problem with the measures of yea-sayers chosen by TER is that TER use the percentage of votes at the highest design point of the first choice question when a second choice question at a higher design point is available. While there are several well-known problems associated with a second dichotomous choice question with a dollar amount contingent on the response to the first question (Alberini, Kanninen, and Carson, 1997), those issues should not affect the proclivity of a respondent to engage in yea-saying. Looking at these estimates for the EVOS Study, the possible range of proportions for yea-sayers drops from 34.2% at \$120 to 8.8% at \$250 (Carson *et al.*, 1992). In the SCB Report (Carson, Hanemann *et al.*, 1994), there is an estimate which (a) uses the responses from a second choice question, (b) incorporates changes made in response to a reconsideration question similar to that in the COS Study, and (c) treats non-taxpayers as votes not for the program (again similar to COS). When that estimate is used, the estimate that TER report of 24.7% at \$215 drops to 7.4% at \$360 (Carson, Hanemann *et al.*, 1994, Table F.10). Second, TER make questionable selections among multiple scenarios in the same study. They use the estimate from the SCB Report for the larger injury scenario rather than that for the smaller injury scenario. Yea-sayers should not be responsive to the scope of the injury any more than they would be to the price. If the smaller scenario is used rather than the larger, the percentage drops to 1.7% at \$360 (Carson, Hanemann *et al.*, 1994, Table F.13). Thus, when looking at the more appropriate measures at the highest design points available, TER's claims seem to have little basis.

The percentage of votes for the program for the smaller injury scenario of the SCB Study and the EVOS Study (Carson, Hanemann *et al.*, 1994) suggests an upper bound on the likely background percentage of yea-sayers of about two percent.[25] Even if we take that upper bound as the percentage as yea-sayers, the impact on the Turnbull is very small. TER calculated the impact of correcting for five percent yea-sayers as a reduction of seven dollars.[26] Thus a correction for two percent would be much less. If the potential yea-sayers are dropped from the sample rather than treated as votes not in favor of the program, the reduction in the Turnbull lower bound mean is even less. It should be noted that the effect of yea-sayers are muted by the properties of the Turnbull estimator. This is not the case with some parametric distributions

[25] It is an upper bound because some respondents *e.g.*, respondents with high incomes, may be willing to pay more than the highest tax amount presented in the survey. Increasing the highest tax amount offered beyond some point would not be helpful because respondents, as noted earlier, would likely begin to perceive the amount as implausible. The SCB Study (Carson, Hanemann *et al.*, 1994) is most relevant for determining the likely upper bound on the fraction of yea-sayers because it has a somewhat higher set of design points, has a smaller injury scenario, and deals with coastal resources and the same California population.

[26] TER's computation method (TER p. 33, footnote 30) appears to be in error as their estimated percentage in favor of the program at \$220 differs from the sample percentage in favor of the program.

where even a small fraction of yea-sayers can have a dramatic effect on estimated WTP.

TER's entire argument here is based on the supposition that the proportion of votes for the program is unexpectedly high; but TER never reveal what they think is the appropriate proportion of affirmative votes at the highest bid point. One may surmise that TER believe that the correct amount is less than 30%, the number of affirmative votes at the highest COS design point. Perhaps TER believe that the prevention of oil spills along the Central Coast of California is not worth $220 so that it is unreasonable for any respondent, much less numerous respondents, to vote for a program at such a cost to them. But since TER are asserting that 30% affirmative votes at $220 is too high, one might expect that contention to be buttressed by some evidence or at least a proposed threshold. Furthermore, we find it surprising that, at one point which we discuss below, TER seriously put forward as the possible range of yea-sayers zero up to the entire proportion of votes for the program at the highest bid point. We do not think most researchers would even suggest that yea-sayers would account for more than 5–10% of *votes for*.

TER offer yea-saying as one possible explanation for these "unexpectedly high numbers of 'yes' responses at high bids". TER then tacitly assume that yea-saying is the cause. But TER explain yea-saying as an anchoring[27] phenomenon rather than as an acquiescence phenomenon. The term *yea-saying* is from the survey research literature on acquiescence (Arndt and Crane, 1975; Couch and Keniston, 1960): under certain conditions, respondents seek the approval of the interviewer by responding in the manner the respondent believes the interviewer will approve. Generally, the literature has assumed that the desirable response to the survey question has been an affirmative response; thus the term *yea-saying*. However, a desire to please the interviewer might also lead a respondent not to vote for higher taxes; so, even if present, the net impact of such acquiescence on the percentage of respondents favoring the plan is unclear. Nonetheless, we will use the term *yea-saying* to mean acquiescence that impels a vote in favor of the program. We tested for possible yea-saying with the EVOS survey instrument using a ballot box so the interviewer did not observe whether or not the respondent favored the program and found no significant difference between the WTP estimates from the treatments with and without use of the ballot box (Carson, Hanemann *et al.*, 1994). While there are some areas such as attitudes toward race in which there appear to be a strong social

[27] As noted earlier, anchoring occurs when an agent conditions their responses on an irrelevant piece of information. An example is the starting point in a bidding game. In a binary discrete choice question, the respondent is told the cost of the program. The respondent should condition their answer on this cost amount. In turn, the statistical procedure used to analyze the data should take this conditioning on the cost amount in its determination of the statistics related to the underlying WTP distribution. The Turnbull estimator correctly accounts for this conditioning on the cost amounts.

desirability impact, the survey research literature suggests that for most topics such effects are small.

The COS Report recounted a test for the presence of *yea-sayers* and *nay-sayers* along the lines of that discussed in Kanninen (1993; published version is 1995) cited by TER. TER dismiss that test; but it is the appropriate test for nay-sayers and yea-sayers if one is prepared to assume a parametric functional form for the latent WTP distribution as TER urge. In a model in which the only covariates are the design points, nay-sayers and yea-sayers would affect the goodness of fit since they are insensitive to the design points. TER do not seem to take this into account in their discussion of the test for yea-sayers. Of the models constructed to test for yea-sayers, TER say the following:

> These models contain no information that would allow the function to differentiate between a rational, plausible "yes" vote and a likely yea-saying vote. [Section 3.4.2]

But the purpose of the test was not to distinguish between yea-sayer votes and other votes; the test was to reveal whether allowing for the possibility of yea-sayers improved the goodness of fit using a common flexible functional form for the WTP distribution. What other information is necessary here? Adding covariates of the type contemplated by TER would have little effect except to dramatically complicate the test. Since yea-sayers are by definition insensitive to price, the test in the COS Report adds a term which allows for the possibility of such a group of respondents and then tests whether the additional term improves the goodness of fit of the model.

In the earlier part of TER's discussion of yea-saying, TER discussed the manner of determining the cost at which the probability of a vote for the program dropped to zero. If a proportion of respondents are insensitive to cost, this probability will drop to zero much more slowly. The test reported in the COS Report adds a parameter which specifies the existence of such a proportion. Since the addition of that parameter does not improve the fit, there is no evidence to believe that there are an appreciable proportion of price-insensitive respondents, including yea-sayers.

Finally, TER find fault with the amount of information provided on the models used to test nay-saying and yea-saying. All the information necessary for replicating these tests is provided in the report. The data for these models are reported in the COS Report (Table 6.3 herein). The models are fairly straightforward to fit. Since there are no additional covariates, one may easily take the data and estimate the models we reported. TER do not suggest what further information they require to replicate our results.

Furthermore, this formal test is not really even needed to show the absence of yea-saying and nay-saying. Because the data do not violate monotonicity, the log likelihood of the Turnbull estimate (–709.48) describes a perfect fit to the data and therefore an upper-bound on the log-likelihood of a model of the data. One does not need to replicate the models we describe to see that the

Box-Cox (LL = −709.63) presented in the COS Report (herein Chapter 5, fn. 21) fits so well that adding an additional term for yea-sayers cannot significantly improve the fit from a statistical perspective. Likewise, the reported log-normal and Weibull spike models also provide a fit that is essentially indistinguishable from that of the Turnbull. These results suggest the absence of any significant amount of yea-saying.

The question that TER should have posed, but did not, is how sensitive these tests are to yea-sayers. Instead TER developed a table which shows the impact of correcting for various percentages of yea-sayers (most of which percentages are improbable). The range displayed in the table covers the range from no yea-sayers to the percentage of votes for the program at the highest design point, *i.e.*, 30%. TER never explicitly mention what percentage of yea-sayers they think might be present. Instead they present their dramatic table which they seem to think justifies the following claim:

> With this extreme sensitivity, the authors need to provide far more justification for their assumption of no effect from yea-saying. [Section 3.4.3]

As noted before, evidence from Carson, Hanemann *et al.* (1994) suggests that the fraction of yea-sayers is likely to be less than 2%. TER might have constructed a table with data points in that range. The table TER provide with a 0–30% range demonstrates a dramatic effect from improbable percentages of yea-sayers but does not provide any resolution within the range of the mostly likely impact on valuation, *i.e.*, if yea-sayers are present in their most likely numbers between zero and five percent. Surprisingly TER suggest that 0–30% is the appropriate range to consider and chides the COS Study for failing to address this problem adequately in light of this purported "extreme" sensitivity which is based on the large impact of an improbably large percentage of yea-sayers.

As to the issue broached above of the sensitivity of the test to the presence of yea-sayers, the test reported in the COS Report would have easily detected the presence of five percent yea-sayers. That test indicates that the percentage of yea-sayers is less than 0.1%, the point estimate for the percentage of yea-sayers.

Finally, TER argue that the Turnbull mean precludes proper treatment of the data by over-extending the analogy that the Turnbull mean is the dichotomous-choice analog to using the simple sample mean from continuous data. TER then cite Carson (1991) on the need to take steps to deal with outliers because they grossly impact the simple means for continuous data and conclude that the Turnbull precludes such steps. TER's pronouncements on dealing with outliers are irrelevant; contrary to TER's assertion, the Turnbull estimator is a robust estimator which is very resistant to the influence of outliers. In particular, it can be seen as a variant of the Windsorized version of the α-trimmed mean discussed by Mitchell and Carson (1989; Carson, 1991) with a very large α.

TER (section 3.5.1) argue that we should not have used the Turnbull estimator when we could generate a proper mean WTP estimate using the Table 6.7 construct validity equation. We found TER's recommended use of the construct validity equation surprising since it should yield a higher mean WTP estimate. However, TER argue that the likely reason we did not use the Table 6.7 equation for this purpose is that we would have had to defend the choice of an upper bound for affirmative votes at the high design point. We consider the Box-Cox construct validity equation in Table 6.7 in considerable detail below.

The Box-Cox equation was an appropriate choice to demonstrate construct validity. TER note that the pseudo-R-square[28] of 0.335 is high for a CV study; in fact, it is very high, not only for a CV study but high for cross-sectional individual level economic studies in general. Regardless of the success of the construct validity equation, any parametric model has the weakness that the estimate of mean willingness to pay is substantially dependent on the distributional assumption made by the analyst. This issue is explored in some detail in Carson and Jeon (2000). We have been concerned with this issue for quite some time (*e.g.*, Carson *et al.*, 1992); and it was in part this dependence on functional form on the part of the parametric and semi-parametric estimators that we used in the past that led us to explore the use of the Turnbull, which needs to make no distributional assumption. Furthermore, this concern is hardly original to us.[29] We should note that while the Box-Cox constant predicts the percentage in favor of the plan at each of the design points quite well, we have no sample-based information on the fit of the Box-Cox at any points other than the design points and using the Box-Cox to predict votes at synthetic design points relies entirely on the functional form of the Box-Cox. Given the impact of the choice of the functional form (see, *e.g.*, Carson and Jeon, 2000), without additional information justifying the adoption of this functional form as a basis for predication, we decline to do so when a superior alternative is available – the Turnbull.

We are surprised that TER take such a strong stand that one should use parametric estimators. Unlike TER's other criticisms that the Turnbull does not allow this or that where we point out that other estimators are available for those purposes, all things considered, it is the Turnbull which is the best estimate of damages. As pointed out in our discussion of the Turnbull estimator, no parametric estimator that is consistent with the observed fractions of the sample responding favorably to the spill prevention program will result in a lower WTP estimate than the Turnbull lower bound. The Turnbull will never overestimate the true mean of the sample, and conservativeness has long been considered a virtue in damage assessment. We would have assumed that TER's clients would have wished us to pursue as conservative an estimation strategy

[28] The definition of the pseudo-R^2 we used is the most widely used definition proposed by McFadden (Maddala, 1983).

[29] For recent work see Haab and McConnell (1997; 2002).

as could be justified. The only possible reason we see for TER's insistence on the use of parametric estimation techniques is that TER believe that a parametric estimate of mean WTP would be extremely sensitive to the distributional assumption made or that any parametric estimate of mean WTP would be so high as to be implausible. This is, however, a trivial point, and one that has nothing to do with contingent valuation data *per se*. Furthermore, as we demonstrate below, it is not even correct. The major problem is extrapolation to the range outside the largest design point coupled with the implausibility of extremely high design points: in fitting parametric distributions to a limited number of design points, one can pick distributional assumptions that can result in radically different estimates of mean WTP. The result of such a choice is that, unlike the Turnbull Lower Bound Mean which will always be a lower bound of the true mean of the sample, the parametric mean can never be separated from its distributional assumption; and its relationship to the true mean is problematic. For instance, assumption of a log-logistic distribution generally leads to the estimate of mean WTP being infinite; yet the number of observations necessary to distinguish a log-logistic distribution from a log-normal one is extremely large. It is always possible that a very tiny fraction of the sample holding very large WTP values can largely determine mean WTP. Likewise, assumptions of distributions such as the normal that impose symmetry restrictions *(e.g.*, the mean and median are forced to be equal) typically result in substantially lower estimates of mean WTP than do parametric functional forms like the Weibull that allow for the possibility of symmetry but do not force it. It is also the case that failure to allow for a flexible enough distribution can result in very bad estimates of mean WTP, particularly if there is any sizeable number of respondents whose WTP for the good is at or very close to $0.

Given the above considerations, if one decides to use a parametric estimator, then one should look at distributions that allow either for a spike at zero or provide substantial flexibility with respect to the shape the latent WTP distribution can take. For instance, if one fits the Box-Cox model with no covariates, the point estimate of mean WTP is $243 with a 95% bootstrap-based confidence interval of $143–$822 using the percentile method (Efron and Tibshirani, 1993). Fitting the Weibull with a spike yields a $285 point estimate of mean WTP, and fitting the more flexible generalized gamma distribution yields a mean WTP estimate of $206. None of these three estimates would be unexpected given the Turnbull lower bound mean of $85. Carson and Jeon (2000) have taken a Bayesian approach to estimating WTP distribution using mixtures of Weibulls. This even more flexible approach approximates a smooth non-parametric solution and generalizes the spike models currently in the literature. Using the data from COS Study, priors that incorporate an income bound on WTP, general knowledge on the shape of WTP distributions, and information from the last COS pilot study, Carson and Jeon (2000) estimate mean WTP at $148 with a 95% confidence interval of $130–$171, an estimate that suggests that the Turnbull lower bound on mean WTP is an underestimate of mean WTP by a factor of about two.

Turning now to results from the Box-Cox construct validity model in Table 6.7 that TER believe we should have used, we find that this model yields a point estimate of mean WTP of $246 with a 95% bootstrap confidence interval of $152–$856, again using the bootstrap percentile method (Efron and Tibshirani, 1993). This estimate is very similar to the $243 mean WTP estimate obtained without using the covariates. Furthermore, this estimate is derived by taking the average of the expected WTP estimates for each respondent in the sample. This is the correct approach to determining the overall mean WTP rather than the TER approach of evaluating the function at the mean value of each of the covariates.[30] In deriving this value, we placed no limits on the upper bound of integration.[31] The apparent explosion in the WTP estimate that TER try to convey will occur as one sets higher levels on the upper bound of integration simply does not occur as one moves this limit upward from $1000. The reason for this is simple: the largest expected WTP value for a respondent is $1184. Indeed, only 10 respondents (less than 1% of the sample) are estimated using the Table 6.7 construct validity equation to have WTP values greater than $1000. This in no way seems implausible.

The issue that TER make about the desirability of basing our prediction on the addition of covariates is *per se* relevant only to predicting outside our design points (and to the demonstration of construct validity). At our design points, no estimator can surpass the Turnbull. A statistical model based upon covariates attempts to optimize its predictions with respect to individual choices, not with respect to being close at each design point.[32] While, as noted earlier, the Table 6.7 construct validity equation gives quite close predictions of the percentage in favor of the program at each of our a design points, this does not in general need to be the case. With covariates, the model attempts to optimize prediction of individual choices rather than necessarily achieving close predictions at each design point.

The TER statement on page 35 that the bootstrap can be used to calculate standard errors for means that reflect the variation in the data is incorrect. The censored data model cannot be calculated via the bootstrap approach

[30] The approach that TER used to look at the upper bound on truncation issue is the wrong way to calculate the means; but without the individual covariate values it was all they could do.

[31] Carson and Jeon (2000) point out such limits may be justified by recourse to economic theory concerning the effective income constraint on WTP values. Even very large bounds on WTP values may substantially reduce estimates of mean WTP in many parametric specifications. This is because in most parametric specifications, mean WTP depends upon both the estimated location parameter and the estimated scale parameter. Treating a large fraction of the data as right censored, as found in many CV studies, including this one, tends to inflate the scale parameter and in turn the estimate of WTP. This is one of the reasons that we are reluctant to rely on the mean WTP estimates derived under the typical parametric assumptions.

[32] Our Table 6.7 construct validity equation does provide fairly good estimates at each design point, although, as expected, not as good as the Turnbull.

without also making a parametric functional form assumption. The bootstrap estimates are conditional both on that parametric functional form and on the design points originally used. It is true that the bootstrap could be used to get confidence intervals for parameters; although these should be reasonably well estimated using standard asymptotic approaches given our large sample size. The bootstrap approach can also be used to look at statistics such as the mean and median WTP that are functions of the estimated parameters.[33] However, nothing in the bootstrap procedure in any way overcomes the fact that the parametric models are making assumptions about what happens outside the observable range of the data.

This same concern motivated our advocacy of the use of the median which TER comment on. As TER state the "median tends to be less sensitive to changes in functional form and other modeling assumptions" (Section 3.5.1). Thus, TER seem to have no trouble understanding why we might use the median. The Turnbull has all the positive qualities of the median and the interval where the median falls can be reliably inferred from it. In addition, the Turnbull estimator will, barring a very poor choice of design points, be closer to the mean than the median. However, the Turnbull mean remains a conservative estimator in that it will always be less than the sample mean. As TER acknowledge in footnote 31, the mean is the appropriate economic measure.

We will continue to use the median when appropriate.[34] We use it in the COS Report for an analysis of sensitivity for the very reasons noted above. One complaint that TER make is that median WTP is only briefly noted. In part this is because it is obvious where median WTP falls by looking at the Turnbull estimator. One can only get a point estimate of median WTP by making a parametric assumption. While it is true that the parametric estimates of median WTP are less sensitive to the particular distribution assumed, it is still that case that, if one is not prepared to make an assumption about that distribution, that the Turnbull interval containing the 50th quantile is all that can be said about the median.

With respect to Hanemann's particular advocacy of the median, Hanemann has always been clear that one might want different statistics for different purposes. In particular, for natural resource damage assessment, N multiplied by mean WTP, or equivalently the sum of WTP over the relevant population (of size N), is appropriate. The public choice criterion with a flat tax is median WTP. For other types of taxes such as an income or property tax, mean of the conditional medians is the appropriate measure. For benefit-cost analysis, the Pareto criteria traditionally calls for mean WTP; although most benefit-cost

[33] For these statistics, TER's argument that the bootstrap is needed is perhaps better founded since these statistics are typically non-linear functions of the original model parameters.

[34] More generally, we believe that a variety of different statistical estimates (and estimators) can be useful in examining the properties of a CV data set.

texts suggest that policymakers do and should consider the entire distribution of benefits.

So we are unable to explain why TER would think our reasons for adopting the Turnbull estimator to be unclear. We can only surmise that TER have failed to separate philosophical arguments about what the correct measure should be and statistical limitations and assumptions that must be made in the process of getting an estimate of the desired measure.

In the next subsection (3.5.2), TER diverge from the discussion of the appropriate estimate to limit the impact of TER's admission regarding the construct validity equation. TER replay their earlier argument that a construct validity equation, even a good one, does not help establish the validity of the estimate of willingness to pay. This is a surprising argument given the widespread acceptance of estimating construct validity by equations in the CV literature (Mitchell and Carson, 1989) and the use TER have made of this technique in their own work, for example, in their widely-distributed monograph prepared for Exxon *Measuring Nonuse Damages Using Contingent Valuation* (Desvousges, Johnson *et al.*, 1992).[35]

We also find TER's criticism (section 3.5.3) of our cluster analysis in section 6.6.2 puzzling. TER appear to believe that we have stated that this cluster analysis demonstrates something about the validity of the WTP estimates that could not be inferred from the construct validity equation (Table 6.7) that immediately proceeds the cluster analysis subsection. We clearly state at the beginning of the cluster analysis section that cluster analysis is an "alternative approach" that is popular in marketing research where the objective is often to segment a population into a small number of subgroups. The results presented in the cluster analysis section suggest that we were successful in doing this with the cluster membership definitions and sizes being quite plausible. However, as we clearly note, there is a loss in information, as would be expected, in moving from 16 predictor variables in Table 6.7 to 3 cluster membership indicators in Table 6.10. We agree with TER's assessment that the cluster membership definitions to some degree capture some of these differences in environmental attitudes. The entire point of the clustering is to capture differences from these and other factors such as demographic characteristics into a small number of cluster membership indicators that represent relatively homogeneous groups. To the extent that these variables are thought to be predictive of differences in underlying WTP, one should expect the Turnbull lower bound mean estimates (Table 6.9) to differ by cluster membership. This is the case (*i.e.*, \$121.53 for Cluster A to \$20.77 for Cluster D).[36] It would be troubling if this result was not observed.

[35] In contrast with the size of the pseudo R-square reported in Table 6.7, the pseudo R-squares reported in Desvousges, Johnson *et al.* (1992) are typically quite small (the pseudo R-squares reported in Tables 5–15 and 5–16 range from .03 to .09).

[36] The cluster analysis approach provides one simple and easy way to interpret the manner the estimate of the Turnbull lower bound mean varies for identifiable subgroups of the population.

Major Flaw 5: Design Flaws that Make the COS Study Almost Useless for Assessing Compensable Value Losses for Oil Spills that Differ from the Spill Described in the Study

The fifth main flaw that TER allege is the presence of "[d]esign flaws that make the COS Study almost useless for assessing compensable value losses for [future California] oil spills that differ from the spill described in the study".[37] TER reiterate their claim that the COS Study estimate is not any good in its own right, but also claims that even if it were, the COS Study provides a single value for a single injury scenario without providing additional information that would allow the estimate to be used in a benefits-transfer exercise to construct an estimate for a different injury.

The details of TER's argument run as follows. First, TER revisit their earlier criticism to re-argue its point that the COS Study does not provide a valid measure for the value of the original COS injury scenario. Second, TER correctly point out that in a benefits-transfer, the flaws of an original study are propagated through a benefits-transfer to the estimate for a different injury scenario. Third, TER correctly point out that the "COS study estimated a single value for a specific scenario with a specific combination of attributes." (p. 40) Fourth, TER state that, to use the COS Study estimate unadjusted, "the services valued in the original study must be sufficiently similar to the services to be valued in the transfer application." (p. 40). Fifth, TER argue that "the likelihood of the injuries in the transfer application being sufficiently similar to the original study is low." Sixth, TER contend the COS Study does not provide sufficient information for adjusting the original estimate to account for differences between the original scenario and the injury scenario to which the estimate is being transferred. TER note a number of different ways that the benefits-transfer injury may differ and thereby identifies the information that they think would be needed to use the COS estimate for benefits-transfers, information which the COS Study does not provide. For example, the level of an individual attribute may change: "The study does not estimate a value for preserving the same number of animals along 20 miles of coast or preserving 20,000 birds ...." (p. 40) TER also note that if the COS injuries doubled, the COS Study provides no information by which the estimate could be adjusted. Furthermore, TER note that the COS Study provides no information on tradeoffs between attributes, so that if two attributes both change but in different directions, information for adjusting the estimate is lacking. Seventh, TER revisit their earlier denunciation of the Turnbull estimator for lacking covariates and contends that the estimate from the COS Study cannot for that reason be adjusted for personal characteristics. Finally, TER contend that it

[37] TER limit this discussion to uses "for estimating natural resource damages from future oil spills in California."

would be inappropriate to transfer the COS estimate to another spill injury that occurred outside the Central Coast.

TER's idealized depiction of a benefits-transfer as starting with multiple estimates scalable by multiple injury attributes has a critical shortcoming. One would think from TER's presentation that benefit-transfers in the past have been rather exacting affairs proceeding from an original study with values for injuries closely matching the injuries of the transfer scenario or with information allowing the scaling of values for the injuries in the original study to the transfer injuries along with information allowing for adjustments due to any differing personal characteristics of the relevant affected population. But TER's position is remarkable: it is a *de facto* condemnation of almost all past benefit transfers. Indeed, TER proclaim criteria that are met by almost no past benefit-transfer exercises, if any. However, TER do not offer any example of a benefits-transfer exercise that would meet this standard. We will reserve for the moment the issue of what sort of study could provide the information that TER require to be available.

Benefit transfer exercises typically take one of two forms. The first type involves an estimate from an existing study that values the benefit of a program that is reasonably similar to the program of current interest. Adjustments are made to that estimate to compensate for obvious differences such as differences in the quantity of the good to be provided or in the income of the relevant population. This type of exercise is often undertaken as a "bounding" exercise to get some idea of the plausible lower or upper bounds on the likely benefits of the proposed program. The second type of benefit transfer exercise takes estimates from a number of different studies and then estimates the benefits of the proposed program as some type of weighted statistical average of the different existing studies used in the exercise. The "weights" are based upon some determination of the relationship between the programs in the existing studies and the proposed program. For both types of benefit-transfers, the basic input is a single estimate for a specific program from an existing study.

The problem with TER's assertion *vis à vis* the COS Study is that at the very least there is ample information available to use the COS Study estimate as an upper or lower bound of the value of a considerable range of injuries as most injury scenarios will be clearly larger or smaller than those in the COS Study across most injury dimensions. The other key issue likely to be a factor is whether endangered species are affected. If endangered species are substantially affected, the COS WTP estimate would likely be a lower bound on the value of the new injuries because the estimate from the COS Study did not include harm to endangered species.

TER also argue that because the COS Study did not value spills on other coastal areas, the COS estimate could not be applied to spills in those areas. This criticism is strange since the estimates used in benefit transfer exercises usually come from studies focused on some other geographic location. For example, Desvouges, Naughton, and Parsons (1992), in a benefit transfer exercise involving water quality benefits, note that "[t]ypically, these estimates [for the benefit transfer exercise] are studies conducted on rivers other than the policy site." They then proceed to use contingent valuation estimates for the Monongahela and Charles River to look at the benefits of water quality improvements on, among others, the Fox River in Wisconsin, the Hudson River in New York, and the Kennebec River in Maine.

The only other explicit point that TER marshal in support of their argument has to do with the purported shortcomings of the Turnbull estimator. These are the same shortcomings which TER argued earlier: since the COS Study Turnbull estimate doesn't have covariates, the estimate cannot be adjusted for respondent characteristics. But this contention about needing covariates is advanced in the wrong context. In its discussion of this flaw, TER specifically note that the area of application we are talking about is estimating the damages from future oil spills in California. The population sampled in the COS Study is the population of English-speaking California households; no adjustment for differences in population characteristics is necessary for any injury scenario as long as one wants to extrapolate to the same California households.

Even though not strictly applicable to this section due to TER's own restriction of this section to California, TER's concern for covariates is appropriate in the larger context. Many benefit transfer exercises are conducted using estimates from a different population in a different geographic area and adjusting the estimate for the characteristics of the population of interest makes the transfer more accurate (Desvousges, Naughton, and Parsons, 1992). Were we to do a benefits transfer in another state requiring an estimate with covariates, we would probably use a flexible model with covariates such as the Box-Cox model presented in Table 6.7 or the Turnbull estimates for each cluster presented in Table 6.9. We also note that one is not constrained to use the set of covariates used in the models reported in Tables 6.7 and 6.9. For instance, if only a subset of covariates used in Tables 6.7 and 6.9 were available for the new geographic area, it would be possible to estimate a model based upon the original COS dataset using only that subset of covariates. Since we discussed the issue of using a Turnbull estimate with covariates earlier, we will not discuss it further here other than to contradict TER's claim that the Turnbull cannot be adjusted.

While TER insist that valuation studies should typically be able to provide values for multiple scenarios, TER refrain from explaining the means by which such estimates might be generated. Because of their advocacy of conjoint analysis (*e.g.*, Mathews *et al.*, 1995; Dunford, Mathews, and Johnson, 1997), we surmise that they implicitly rely on the availability of conjoint analysis to provide such multiple values for a benefit-transfer, at least where passive use values are involved. Consequently we examine the viability of conjoint analysis both as a means of fulfilling this role and as a substitute for the standard type of contingent valuation study.

Conjoint analysis has a substantive role to play in environmental valuation.

Indeed, two of us were involved in one of the first modern style choice-based conjoint analysis studies in environmental economics (Carson, Hanemann, and Steinberg, 1990). Not only have we conducted many conjoint analyses, but one of us coauthored one of the standard review articles on choice-based conjoint analysis (Carson, Louviere *et al.*, 1994). The problem with conjoint analysis that we raise here lies not with conjoint analysis *per se*, but in the claims TER make for it and in TER's failure to recognize key problems in its use in the valuation of environmental public goods. The properties of conjoint analysis have been little studied with respect to public goods; and the first published conjoint analysis of a public environmental good – a woodland caribou management program in Alberta Canada (Adamowicz *et al.*, 1998) – only appeared three years after the COS Study was completed.

While sometimes touted as an alternative to standard CV, conjoint analysis is a simply one variant of CV (Adamowicz *et al.*, 1998). The two specific forms of conjoint analysis that have been used in the environmental economics literature simply generalize the binary discrete-choice elicitation format used in standard CV surveys like the COS survey by either offering respondents more than two choices, *i.e.*, a multinomial choice (Adamowicz, Louviere, and Williams, 1994), or by asking respondents a sequence of paired comparisons (Johnson and Desvousges, 1997).[38] None of the relevant issues that TER raise with respect to standard CV can be eliminated by moving to either of these two elicitation formats. In short, conjoint analysis is heir to all the cognizable issues that should be of concern in doing the standard variant of CV, exemplified by the COS Study. While many of the issues TER raise in this Critique are irrelevant to CV in general, many merely miss their mark in regard to the COS Study and are important considerations in doing contingent valuation. These issues alone should give TER pause in its *sub voce* reliance on conjoint analysis.[39]

However, both the two conjoint analysis elicitation formats mentioned above – multinomial choice and sequence of paired comparisons – pose additional problems relative to a standard binary discrete choice CV survey.[40] First, a

[38] The paired comparisons typically take the form of binary discrete choice questions. In some instances an effort is undertaken to get the respondent to provide an indication of how much more one alternative is favored. The use of this *extra* information beyond that in the binary discrete choice question requires assumptions about agent utility that go beyond that assumed by the standard neoclassical economic model and hence are avoided by most economists.

[39] We should reiterate that TER do not invoke conjoint analysis explicitly but rather by explicitly relying on end-products that could only be produced by conjoint analysis.

[40] In the marketing literature, respondents are sometimes asked to completely rank order a set of alternatives. This practice appears to be declining, in part, because choice-based questions are easier for respondents to answer and, in part, because of demonstrations that the typically used statistical approaches for analyzing rank order conjoint data require the assumption that rank ordering produces the same result as asking individual questions asking about all possible binary discrete choice pairs of alternatives (Chapman and Staelin, 1982; Carson, Louviere *et al.*, 1994)

binary discrete choice question in a context such as that of COS scenario (*i.e.*, a public good with a coercive tax payment mechanism) has incentives for truthful preference revelation (Carson, Groves and Machina, 1999).[41] When one moves away from that elicitation format either by asking respondents about more than two choices or by offering respondents more than one choice scenario, truthful preference revelation is no longer always the optimal response strategy for an agent (Carson, Groves, and Machina, 1999).

Second, the presentation of more than two choice alternatives in the survey, of necessity, implies that a less detailed presentation must on average be made for each of the alternatives offered. While there may be some economies of scale in presenting multiple alternative programs, for a fixed length survey, the average amount of time devoted to each specific alternative must of necessity decline. This decline in the amount of information provided on each scenario tends to be very substantial except in surveys involving a small number of very tightly linked goods.

Third, the analysis of conjoint data, at present, generally requires the invocation of strong parametric modeling assumptions (rather than non-parametric and robust estimators like the Turnbull estimator); these assumptions exert such a large influence on the constructed estimate that the choice of assumptions becomes an unresolvable issue of controversy in itself.

While a conjoint analysis survey can collect respondent preferences involving a substantially greater array of alternative policies than a CV survey of the standard variant, there are clear costs involved in such a tradeoff. TER do not seem to appreciate these costs.

Major Flaw 6: Design Flaws that Make the COS Study Useless for Scaling Compensatory Restoration Options Under the New NOAA Natural Resource Damage Assessment (NRDA) Regulations

The last of TER's purported flaws is that the estimate from the COS Study cannot be used under NOAA's new NRDA regulations. As with the benefits-transfer section above, TER recapitulate their prior criticisms in a slightly different context; and as with the benefits-transfer section, the first part of this section of TER's Critique is less a discussion of additional flaws than a contention that the previously-discussed flaws make the study unuseable in some additional way, here for damage assessments following NOAA regulations.

[41] The aspect of public goods that is important here is that all agents share the same level of provision of the public goods. This is fundamentally different than the situation involving private and quasi-public goods where agents get to determine their own individual levels of consumption. As such the performance characteristics of conjoint analysis with respect to private and quasi-public goods may be considerably different than it is with respect to public goods.

TER discuss service-to-service scaling only long enough to glibly assert that the COS Study could not be used in service-to-service scaling. This point seems superfluous given that if service-to-service scaling is the chosen for its common metric, any constructed estimate of monetized value would be useless since service-to-service scaling makes no use of monetized values. Since the COS Study estimate is in dollars, no direct service-to-service scaling is possible. TER imply that CV *must* involve a monetarized tradeoff, implying that service-to-service scaling must be done with some unspecified method, a method which we presume to be conjoint analysis. However, it is certainly possible to ask respondents a binary discrete choice CV that offers a service-to-service tradeoff which does not involve money.[42] As with standard CV, we presume that conjoint analysis could also present a tradeoff of two similar goods with the same metric. Whether the metrics of the respective goods in the tradeoff made in an economic choice are dollars and wetland acreage or both goods are measured in wetland acreage is irrelevant to the principles involved in presenting the tradeoff to respondents. Thus the only advantage of conjoint analysis for service-to-service scaling lies in its ability to collect information from each respondent about multiple programs. We have already noted, however, this advantage is gained by moving away offering a single binary choice and that move entails certain costs.

In addition, the service-to-service approach may be theoretically unsuitable for natural resource damage assessment. While it is possible to ask survey questions intended for this purpose, and tradeoffs between program attributes can be well-defined from an individual agent's perspective, aggregate measures of compensation have recently been shown not to be consistently defined in standard economic welfare terms unless they are based on a common money metric numeraire (Flores and Thacher, 2002). The intuition behind this result is straightforward: since the compensatory good to be provided is a public good, only one specific level of the good can be provided to all agents. Hence it is not possible to make agents *whole* on an individual-by-individual basis. It is impossible to determine how to compare the gains and losses of different agents with respect to the original injuries to the public resource and the purposed level of provision for another (compensating) resource unless one assumes a form of *cardinal utility* specified in terms of the two public resources being considered. Since economists have long rejected making the cardinal utility assumption (see, *e.g.*, Just, Hueth, Schmitz, 1982.; Varian, 1992) the only solution is to use a money metric (*i.e.*, valuation scaling).

[42] For instance, Shyamsundar and Kramer (1996) use baskets of rice in Madagascar in an area where the cash economy is very limited. Monetarized tradeoffs are more often used in CV because they usually the most useful estimates. All the comparison modes commonly used in conjoint analysis studies, including the paired comparison favored by TER (Johnson and Desvousges, 1997), may also be used in CV studies.

TER next claim that the COS Study is also useless for valuation scaling. TER describe the valuation scaling approach as follows:

> Using the valuation scaling approach, the value of the lost interim services is determined, and then a compensator-restoration plan is devised that will generate services of equal value to the public. Thus, valuation occurs for both the forgone services as a result of the spill and the increase in services as a result of the compensatory-restoration actions. [Section 4.2]

TER then seem to distinguish between the valuation of the lost interim services and the valuation of the compensatory services. This distinction without a difference seems to arise only because TER allow that the first type – valuation of lost interim services – might be valued by a standard CV survey with a single scenario; while TER claim that the second type – the valuation of compensatory services – could not because "[i]t is unlikely that a restoration alternative would be devised that provides the same services as described in the COS scenario for preventing an oil spill." Thus the distinction posed by TER is not a real difference between two types of scenarios; rather it is that same benefits-won't-transfer conclusion from earlier that a study that values only a single scenario cannot provide the multiple estimates for benefit-transfers to other goods, in this case for valuing compensatory services. More specifically, TER say that the COS Study could never be used "to value restoration alternatives because values are not obtained for specific services that would result from different restoration alternatives." TER's point is illustrated by its example that follows:

> In addition [to valuing the lost interim use], it would be necessary to determine the appropriate scale of possible alternative [*sic*] for replacing the services lost as a result of the spill's injuries based on the value of the services provided by the alternative. For example, one possible compensatory-restoration alternative might be to expand wetlands that provide food and serve as nesting area for particular bird species. Another alternative might be to modify rocky intertidal areas in order to enhance the recruitment and growth of benthos in that ecosystem. Yet another alternative might be a combination of the first two alternatives. [Section 4.2]

Thus TER assume that a single survey could cost-efficiently produce multiple estimates of alternative restoration scenarios. This is also the linchpin of TER's benefits-transfer argument; and our response to its use here is the same. Anything that can be done with a conjoint analysis study can be done with several binary discrete choice CV studies. All things considered, as input for a benefits transfer or for the valuation of compensatory services, the COS Study will be as useful as any other study. Thus we believe that single scenario CV study can provide a reliable starting place for comparison to other resource-based scenarios, including compensatory restoration scenarios.

Conclusion

TER have two basic claims: first, that the estimate from the COS Study value is not reliable; and, second, that the COS estimate is not transferable to the set of injuries that might occur from an actual oil spill. In regard to the details of the first claim, we have refuted each of TER's assertions about flaws that go to the reliability of the COS Study. In regard to the second claim, the advantages of a study to estimate willingness to pay to avoid damages for a stylized set of oil spill injuries may be viewed from five distinct perspectives: first, consideration of public policies that may influence oil spills; second, provision of information to those shipping oil on the magnitude of likely damages; third, provision of initial estimates for discussions of compensation from a spill; fourth, provision of a substantial input to a benefit transfer study looking at the damages from an actual spill; and fifth, provision of much of the basis for a new CV study focused on an actual set of injuries, if it becomes desirable to conduct such a study. We find that the TER's Critique on the fourth of these, the benefit transfer potential of the COS Study, is without merit. As a very high quality CV study dealing with a set of injuries that have to be clearly related to the set of injuries that would occur from any major oil spill occurring along California's Central Coast, it is inconceivable that any legitimate analyst doing a benefit-transfer exercise for the oil spill would not heavily rely upon the COS Study. TER have postulated the existence of standards that are far beyond the current practice of benefit-transfer (including those TER have carried out), standards that would preclude the use of almost every valuation study ever conducted. The apparent subtext behind TER's Critique is that a conjoint analysis study could provide what TER claim the COS Study cannot. However, conjoint analysis is simply a variant of CV that asks each respondent about multiple scenarios. In addition to the considerations to which conjoint analysis shares with the standard variant of CV, the use of any of the commonly used conjoint analysis elicitation formats to value public goods raises a number of additional serious issues. Not only does TER never confront these issues in their Critique of the COS Study, they never even identify conjoint analysis as the source of their multiple estimates.

Index

The Economics of Non-Market Goods and Resources

1. J. Loomis and G. Helfand: *Environmental Policy Analysis for Decision Making.* 2001 ISBN 0-7923-6500-3
2. L. Fernandez and R.T Carson (eds.): *Both Sides of the Border.* Transboundary Environmental Management Issues Facing Mexico and the United States. 2002 ISBN 1-4020-7126-4
3. P.A. Champ, K.J. Boyle and T.C. Brown (eds.): *A Primer on Nonmarket Valuation.* 2003 ISBN 0-7923-6498-8
4. P. Dasgupta and K.-G. Mäler (eds.): *The Economics of Non-Convex Ecosystems.* 2004 ISBN Hb 1-4020-1945-9; ISBN Pb 1-4020-1864-9
5. R.T. Carson, M.B. Conaway, W.M. Hanemann, J.A. Krosnick, R.C. Mitchell and S. Presser (eds.): *Valuing Oil Spill Prevention.* A Case Study of California's Central Coast. 2004 ISBN 1-4020-2326-X

KLUWER ACADEMIC PUBLISHERS – DORDRECHT / BOSTON / LONDON

GPSR Compliance

The European Union's (EU) General Product Safety Regulation (GPSR) is a set of rules that requires consumer products to be safe and our obligations to ensure this.

If you have any concerns about our products, you can contact us on ProductSafety@springernature.com

In case Publisher is established outside the EU, the EU authorized representative is:

Springer Nature Customer Service Center GmbH
Europaplatz 3
69115 Heidelberg, Germany

Batch number: 10725825

Printed by Printforce, the Netherlands